Ian Breach

Windscale Fallout

*A Primer for the Age of
Nuclear Controversy*

Penguin Books

Penguin Books Ltd, Harmondsworth,
Middlesex, England
Penguin Books, 625 Madison Avenue,
New York, New York 10022, U.S.A.
Penguin Books Australia Ltd, Ringwood,
Victoria, Australia
Penguin Books Canada Ltd, 2801 John Street,
Markham, Ontario, Canada L3R 1B4
Penguin Books (N.Z.) Ltd, 182–190 Wairau Road,
Auckland 10, New Zealand

First published 1978

Made and printed in Great Britain by
Richard Clay (The Chaucer Press) Ltd, Bungay, Suffolk
Set in Monotype Plantin

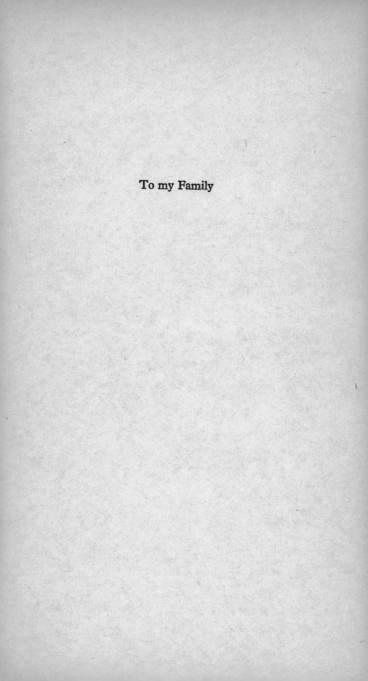

To my Family

Contents

Contents

Acknowledgements

My thanks are due to many people who, in one way or another, made possible the writing of this book. Among them I should include J. D. F. Jones and David Fishlock, respectively managing editor and science editor of the *Financial Times*; also Bernard Dixon and Michael Kenward, editor and news editor of the *New Scientist*. Without the assignments from these two publications to cover the Windscale inquiry hearings through the summer and autumn of 1977, my acquisition of the necessary raw material would have been a very much more difficult and expensive business.

Naturally enough, the parties to the inquiry have, when approached, been invariably ready to supply additional information to help clarify points of detail. I record my gratitude nonetheless, for *Fallout* is largely a gathering-together and interpretation of *their* proposals and ideas. Other and deeper analyses of the nuclear-power controversy are sure to appear in due course. In producing this highly compressed account, my aim was to provide a ready-reckoner for the debate, with Windscale to be seen as an episode rather than as a whole story.

As a piece of reprocessing, its faults will be all too readily discerned by those who have contested the issues over the past two years or so. Its technical satisfactoriness may be challenged; its economic benefits are dubious; and its employment potential has been limited. On the other hand, it poses no threat to civil liberties, its biodegradable, and presents no hazard to environmental or occupational health and safety. As for proliferation, the only kind I hope for is that of new ideas.

Milburn, May 1978

1 Introduction

On 15 May 1978 the British Parliament made a momentous decision. By a majority of 144 votes, it ratified the Government's granting of permission to British Nuclear Fuels Limited to construct a plant for the reprocessing of spent fuel from nuclear power stations in the UK and overseas. The reverberations of that decision will be felt by millions of people around the world and in many aspects of their lives. Its consequences will last for longer than the recorded history of our species. This book attempts to explain the decision and the events that have led up to it, including the Windscale inquiry – the longest nonstop exercise in public participation yet held in England.

The controversy over whether a reprocessing plant should be built at Windscale or indeed anywhere else has become the arena for fundamentally and intensely differing views of the way the modern industrial state should develop. In both its conclusions and the way in which they were expressed, the official report on the Windscale inquiry, written by its chairman Justice Parker, is a focus for the controversy. It has been accepted by the Government, approved by the House of Commons and applauded in most of the press. It has also been received and read with disbelief and anger, frustration and condemnation by scientists, lawyers, churchmen and members of the UN, and by almost the entire environmentalist movement.

Since this account of the Windscale decision and its implications is aimed at – and, I hope, will reach – a much wider audience than those who have been involved with or have closely followed the affair, let me begin with a few excerpts – the kind of comment to which readers will have been exposed – from the British press on publication of the Parker report – starting with two extracts from 'popular' newspapers, the *Daily Express* and the *Daily Mail*. The *Express*, in a leading article entitled 'Safer Than You Think', had this to say:

Parliament will shortly debate a report which approves the establishment of a nuclear reprocessing plant at Windscale.

Clearly Peter Shore, the Secretary for the Environment, agrees with the proposal and he is absolutely right.

Emotive phrases like 'nuclear dustbin' should not disguise the fact that nuclear power has so far proved infinitely safer than most of heavy industry.

One wouldn't think so to hear the boys on the outrage circuit, but nuclear energy has not taken a toll of human life in this country. That is more than can be said about coal mining, deep sea fishing, or asbestos working.

A distinction must be made, however, between the risk of an accidental leakage, which is infinitesimal, and the chance of plutonium being stolen by the sort of people who hijack aeroplanes.

One fact that must have weighed with Mr Shore is that plutonium is in safer hands in this country with its police and Army than in any other country.

To be frank, we need the money. If Britain had made a success of its heavy industry, maybe it could turn up its nose at the chance of £900 million worth of business. But it hasn't and it can't.

And the *Mail*:

... Opponents of the plan to reprocess British and foreign nuclear waste up at this site in Cumbria may object to the findings of the report published yesterday. They cannot reasonably complain that their case has not been given a full and sympathetic hearing.

Not only have we had a full-blown inquiry, but Parliament also is now being invited to have its say as well.

... Our hunch ... is that this most controversial development at Windscale should be given the go-ahead.

The *Daily Telegraph*, less inclined to the cosy-chat approach to leading articles but more or less politically aligned with the first two, wagged a stern and chauvinistic finger while approving the report:

Some of the objections to nuclear power could apply to almost any manifestation of an advanced society. It was said, for instance, that terrorists might seize some of the nuclear material. So they might; just as they seize aeroplanes. The remedy here is in stronger anti-terrorist measures, not in the blunting of industrial progress ... Nuclear power is potentially cheap and [a] capital-intensive form of energy ... It is thus less vulnerable to being cut off by politically motivated trade unionists or (as in the case of oil) foreign powers. For these reasons, we must hope it becomes the energy of the future.

The *Guardian*, like all the others, saw the report in the late afternoon of 6 March and had no more than two hours in which to make up its mind for the following morning's issue on the Windscale report; and it liked most of what its leader-writer read. But he or she did at least countenance further genuine debate:

The inspector accepts, and it will be interesting to hear the arguments against his view, that the dangers both of proliferation of nuclear weapons and of terrorism can be overcome by supplying the resulting plutonium, after reprocessing, in irradiated fuel rods which cannot, without great hazard and much time, be used for any other purpose.

The report, thought the *Guardian*, would give people a deeper knowledge of what they were embarking upon than in countries 'which have decided to skip the democratic process ... To that extent, Mr Justice Parker's report ... could stand translation into several European languages.'

Only *The Times* expressed a strong reservation. Although it declared itself to be unsurprised by the Parker conclusions, it felt that there had to be strong misgivings over the report's handling of international issues:

There is a strong commercial incentive to take in other nations' nuclear washing while doing our own, and a corresponding temptation on British Nuclear Fuels to make light of the objections [but] if reprocessed plutonium is handed back to other countries at present without nuclear weapons, those countries are brought nearer to military nuclear capability. This possibility has led President Carter to urge other countries not to distribute plutonium in this way.

The Times had other qualifications, which set it aside from its sister papers in London, on the conclusions reached by Parker. But that, as far as British editorial opinion was then concerned, was that: the issue had been raised, debated and disposed of.

Within days, however, it was clear that a good many people thought otherwise, and a stream of impassioned letters began to play upon the *Guardian* and *Times*, the only two British daily newspapers regularly to publish communications from the environmentalist movement. The message of these letters, broadly, was that the objecting evidence at the Windscale inquiry had been ignored or, at best, misunderstood. Professors wrote of their incredulity at the findings of the Parker report; environmentalists alleged that their submissions to the inquiry had been dismissed out of hand. The report, wrote one *Guardian* correspondent,

is without doubt an insult to the integrity and competence of the environmental movement . . . [and] a dangerous document. It gives a stamp of approval to the nuclear programme and placates the fears of those who are perhaps unaware of the scale of economic, environmental, and social dislocation a commitment to nuclear power is likely to bring . . . The environmental lobby yet again has been taken for an expensive ride. It has been used as a puppet in the display of democratic decision-making and as a foil for the cut and thrust of 'progress'.

Journals which had a little more time to digest the Parker report were no less critical. The science periodicals *Nature* and *New Scientist* both expressed strong feelings about the document: the *New Scientist* accused the judge of misrepresentation; *Nature* observed that 'it does somehow seem as if, once Mr Parker had decided that the decision should go in BNFL's favour, he went out of his way to find for them on almost every issue – rather as a judge, confronted by a bunch of witnesses prepared to testify to a man's innocence, might dismiss their evidence *in toto* once satisfied that the man was guilty.' The *New York Times* went straight to what its readers would regard as the point: 'Britain is about to undermine President Carter's campaign to curb the spread of nuclear technologies that can be used to make bombs. A British judge has recommended construction of a nuclear fuel reprocessing plant at Windscale, on England's north-west coast.'

And the London *Observer*, as consistently heterodox on nuclear matters as it has been over the years on supersonic transport, reinforced the American position in its leader on the following Sunday:

MPs should resist pressure to hurry a decision when they debate Mr Justice Parker's Windscale report ... the business is uncertain anyway, because the United States, which supplies [nuclear] fuel to Japan, can forbid its reprocessing ... A two-year programme of top-level international meetings began last year, after President Carter's decision to halt reprocessing in the US, with the object of finding nuclear technologies which do not risk proliferation; Britain and Japan are co-chairmen of the group discussing reprocessing and the fate of plutonium. We should, both in prudence and good faith, at least wait to see what emerges from these talks.

The Town and Country Planning Association's director, David Hall, responded to a comment in the *Financial Times*, where it was suggested that the report be sent to America as part of the Government's contribution to the 'International Nuclear Fuel-Cycle Evaluation programme' (INFCE) set up in 1977 by President Carter. So it should, said Hall in reply: instead of hearing second-hand praise of it, the Americans would be able to examine it in all its shakiness.

Then came – and they are still coming as this is written – the more elaborate and detailed criticisms of Parker. Several witnesses at the Windscale inquiry indicated that they were contemplating legal action in view of the serious misconstructions that he had put upon their evidence. Two of the environmentalist groups demanded apologies. Four of them published extensive commentaries on the inquiry and the Parker report, cataloguing errors of omission and commission. A letter from one group, the Network for Nuclear Concern, to the Environment Secretary is typical. It ends with these words:

... we carefully followed the submissions of other objectors at the inquiry and can testify that the misrepresentations we have detailed are merely an example of the selective and inaccurate way in which the issues have been treated in the report.

Finally, as the inquiry report will be taken by some as an objective statement of the facts, we are concerned lest it unjustifiably discredits environmental groups such as ours, who put a carefully reasoned case, as being nothing but alarmists. We therefore urge you, as a contribution to natural justice, to set the record straight and issue a public apology and amendment to the report.

A few weeks earlier, the man to whom this letter was addressed had told Parliament that the Parker report 'analyses in a masterly way' all the issues raised at the Windscale inquiry and that it reached conclusions which he found 'persuasive and broadly acceptable'. His Party leader and Prime Minister, James Callaghan, also approved of the report: he found it 'most cogently argued'.

When Members of Parliament assembled in the House of Commons on 15 May 1978, they were asked to consider the Liberals' motion that the Windscale Special Development Order – approving the construction of a new nuclear-fuel reprocessing plant – be withdrawn. At the outset they were effectively told that the controversy had run its course. The Deputy Speaker opened by saying: 'This is the third time that this subject has been debated in the House, and therefore I feel I have every right to ask hon. Members who have spoken previously on this subject to make very brief speeches if they are called.' Three hours later, the Secretary of State for Energy, Tony Benn, called upon Parliament – or that half of it that was present – to pass the Order. By a majority of 224 votes to 80, they did.

For them, as for British Nuclear Fuels Limited and its industrial and government associates throughout the world, Parker was greeted as the end of a tiresome but tolerable debate. Now they could get on with the business of business, unhampered by antinuclear groups, ecologists (whatever *they* might be), peace lobbies and other opponents of their energy strategies and development plans. But what they were seeing was simply the first phase of a conflict that now seems certain to embroil them and their countries for a very long time to come. More importantly and ominously, it is a conflict in which there can be no winner on present reckoning. The main protagonists can achieve their ends only at the expense of peace or freedom – or both.

2 Who Needs Nuclear Fuel?

Though the Parker report on the Windscale inquiry is proving to be one of the most contentious ever presented to a Secretary of State, it is helpful to opponents, proponents and the general reader alike in its lucid setting-out of the technical background to nuclear-fuel reprocessing. The facts have appeared elsewhere, but the report brings them together in a convenient fashion which, in part, it is appropriate to follow here.

Commercial nuclear reactors are fuelled with uranium, extracted from an ore known commonly as 'yellowcake', the main deposits of which, outside the communist world, have been found in the USA, Canada, Australia, South Africa and Namibia. Refined, natural uranium consists mainly of the isotope U_{238}; a very small proportion – about 0·7 per cent – is of the fissile isotope U_{235}, a form which, when bombarded by neutrons, divides atomically and releases energy in the form of heat. U_{238}, though not itself fissile, will – if it absorbs the 'spare' neutrons discharged when U_{235} undergoes fission – rapidly change into another element, the fissile isotope of plutonium, Pu_{239}. This is one of the so-called transuranics – substances with atomic weights beyond that of uranium and, on earth, usually synthetic. Several of these elements – of atomic numbers 93 and above – are produced during the fission process and reside in the spent fuel when it is removed from the nuclear reactor.

In Britain, two types of reactor are in current commercial use. They are known respectively as Magnox reactors and advanced gas-cooled reactors (AGRs). Both use uranium as a fuel, assembled in such a quantity and disposition as to start the fission process. The speed and intensity with which this takes place is controlled or 'moderated' by the use of neutron-absorbing materials such as carbon or boron. The fuel is left in the reactor for its 'burn-up' life – an optimal period of five years or more chosen to give the maximum amount of heat for the longest possible time without so highly irradiating the fuel as to make subsequent removal and treatment intolerably costly and hazardous.

Magnox reactors burn natural uranium metal, which is prepared in the form of rods and encased or clad in a *magn*esium *ox*ide alloy – the source of their name.

AGRs use a fuel containing an unnaturally high proportion of U_{235} – known as enriched uranium. Uranium metal, refined at BNFL's Springfields works, is enriched at Capenhurst to three or four times the natural U_{235} content, prepared as uranium oxide, and formed into fuel pellets which are then encased in stainless steel, capable of withstanding the greater operating temperatures of AGRs. This is the 'thermal-oxide' fuel which, when burned up, would be reprocessed in the thermal-oxide reprocessing plant (THORP) which BNFL have applied for permission to construct.

Britain has fourteen nuclear commercial power stations, nine of which are based on Magnox reactors, the first of which – at Berkeley in Gloucestershire and Bradwell in Essex – were commissioned in 1962. Of the five AGRs, two – at Heysham in Lancashire and Hartlepool in Cleveland – were uncompleted in 1978 and both running behind construction schedules as a result of industrial action and technical difficulties. The other three – Hinkley Point in Somerset, Hunterston in Ayrshire and Dungeness in Kent – have all produced power but have also been the subject of grave problems: large cost overruns and serious design and operational faults. In England and Wales, the capability from nuclear plants is roughly 6 per cent of the countries' total electrical-power generation. In Scotland, the nuclear capacity is more than twice that figure – making it the world's leading civil nuclear power. If the Generating Board's plans were to be fulfilled for present stations, the capacities would rise, by 1980 or thereabouts, to 20 and 30 per cent, respectively.

There are a number of other thermal-reactor types in use, either in prototype in Britain – like the steam-generating heavy-water reactor (SGHWR) – or in commercial use in other countries, such as the light-water reactor (LWR). Fuel from LWRs would be reprocessed in the Windscale THORP. Since LWRs run at lower temperatures than AGRs, the fuel does not need to be clad in stainless steel; instead, it is encased in a zirconium alloy (Zircalloy).

When fuel is removed from a thermal reactor, of whatever type, its contents are complex but consist basically of three parts: a very high proportion of uranium (about 97 per cent, including residual quantities of the isotopic form U_{235}); between 0·1 per cent and 1 per cent of plutonium; and 2–3 per cent of fission by-products – heavy-atom substances formed by neutron absorption. These are known as actinides.

The only other type of reactor with which the Windscale inquiry concerned itself is the non-thermal fast breeder – the FBR. None is yet in commercial use anywhere, but several countries have built and operated prototype or demonstration FBRs. The UK's are at Dounreay, in north-east Scotland – one of the remotest points in mainland Britain. The FBR is fuelled with a mixture of plutonium and $uranium_{238}$ and, during the fission process cooled by highly volatile liquid sodium, breeds additional plutonium. Over twenty-five years, an FBR would – if working to specification – produce enough plutonium to fuel another FBR. The demand for U_{238} would be so reduced, under ideal conditions, as to stretch nuclear-fuel supplies some fifty to sixty times as far as if they were used in thermal reactors. The technology is fraught with uncertainty, however, and afflicted – as we see later – by the political, social and environmental problems associated with a 'plutonium economy'.

Radioactive substances are unstable: they continuously discharge or radiate particles or energy quanta and 'decay' to progressively lower isotopic states until, eventually, they revert to a stable element. Depending upon the element concerned, that decay can take minutes or it can take aeons. Throughout that time, it is emitting radiation, which may damage living tissue in various ways. Three principal forms of radiation are involved – alpha, beta and gamma. Alpha radiation is emitted only by heavy-atom elements and can penetrate water or soft tissues by, at the most, a few hundredths of a millimetre. Plutonium, for instance, can be safely handled with no protection more than a pair of surgical gloves. Its toxicity lies in its effect if the particulate form is inhaled even in minuscule quantities, when it will eventually kill. Beta radiation can penetrate tissue by a centimetre or two; gamma radiation by several tens of centimetres.

The damage potential of radioactivity will depend partly on the organs or tissues that receive radiation doses, the time over which the doses are received and the source of the radiation, which can span a range from the almost immeasurably mild to the unbearably pernicious; it would also depend upon whether the organism was subjected to an external source of radiation or whether – in the case of an animal such as man – a radionuclide (radioactive substance) had been ingested by inhalation or by way of food or drink. Two main effects will follow doses above a certain level – somatic damage, in which the most common form would be a malignancy or other kind of cancer; and genetic disruption, in which the mechanisms of growth and reproduction would be harmed.

Radiological-protection standards have been researched and laid down for all the known radionuclides, with maximum permitted doses set for workers in the nuclear industry as well as for members of the general public. Recommendations are also in use for the maximum annual exposure that should be received by those belonging to 'critical groups'. The standards include maximum levels for the environmental transmission of radionuclides along *recognized* pathways of the ecosystem, such as food chains and marine or terrestrial habitats.

Spent fuel is highly radioactive and generates large quantities of heat – and would do, if left alone, for a long time after it was discharged from a reactor. Initially, it is stored in ponds by the reactor sites until it is thermally and radiologically cool enough to permit handling under suitable shielding without undue hazard or complication. It is then, in the UK, transported in specially designed flasks, usually by rail but occasionally by road in armoured transporters, to Windscale, where it is stored for a further period in cooling ponds. The THORP complex would deal with spent fuels from AGRs and LWRs in what would be a technological extension of the present Magnox reprocessing facility at Windscale.

Magnox spent fuel is kept for only a short time in the Windscale cooling ponds: the fuel cladding is easily corroded and would, if punctured, allow an unacceptable escape of radioactivity to the pond water. After the cooling period, the fuel rods are removed from the cans, stripped of their cladding and chemically processed. From this, the following products are thus isolated:

* Most of the uranium is retrieved and stored for re-use in existing types of reactor.
* Almost all the plutonium is separated and stored for use in mixed fuels for existing reactors or possible use in commercial fast breeders.
* Highly active fission products, together with residues of uranium and plutonium and other actinides, are combined into a single liquid effluent and stored in shielded, cooled tanks known as highly active waste tanks (HAWs).
* Low- and intermediate-active wastes, in solid, gaseous and liquid form, are separated, of which –

   low-level liquid waste is discharged to the Irish Sea through a shore-line pipe terminating 2·5 km off the coast;

   low-level gaseous waste is vented to the atmosphere through stacks;

   low-level solid waste (including lightly contaminated office and works equipment) is buried in trenches at a site – Drigg – near the plant;

   intermediate solid and liquid wastes are stored against 'ultimate' disposal, possibly at sea if international agreements can be secured.

It is BNFL's intention that the contents of the HAWs, massive and immensely costly structures, should eventually be solidified by being mixed with a borosilicate glass and vitrified. The resulting blocks of waste would be further clad and then removed for long-term storage in deep-mined strata or beneath the ocean bed. Development of the process, known as

HARVEST (after relevant work at Harwell), began in the late fifties but has only recently been the subject of a large research and development programme. A full-scale plant to demonstrate HARVEST is expected to be working some time in the eighties at Windscale: outline planning permission for the plant was granted in March 1977.

Metal fuels have been reprocessed at Windscale since 1952 – initially to produce plutonium for the British nuclear-weapons programme. The operations are regarded as having been successful except in two respects. There have been corrosion problems with Magnox cladding: since 1970, greatly increased levels of caesium contamination in the cooling-pond water – and thence the Irish Sea – have resulted. There have also been difficulties in the first stage of reprocessing – stripping fuel of its cladding. Technical problems have been aggravated by industrial action, with the result that ponds have contained more and older fuel than they should. The company has been given permission in outline to provide new stripping plant and pond-water treatment facilities to overcome these difficulties.

In its reprocessing history, the Windscale plant has handled some 19,000 tonnes of spent nuclear fuel. From this, about 10 tonnes of plutonium has been separated, of which about 7·5 tonnes remains in store. Further reprocessing of Magnox fuels is estimated to yield a total of 45 tonnes of plutonium by the end of the century. Opponents of the THORP scheme have largely accepted that Magnox reprocessing should be allowed to continue. Thus, whatever is done about other thermal-reactor fuels, there will, by the year 2000, be more than 50 tonnes of plutonium recovered from Magnox fuels, plus large quantities of highly active waste – kept in HAWs if the HARVEST process has not by then been perfected and brought into use. Parker concluded, in spite of the absence of hard proof, that the process *would* be successfully established, adding: 'Indeed, success must, either by the HARVEST process or some other process, be achieved.' He did not consider, since it was not canvassed during the inquiry, alternative means of disposing of highly active wastes – one of which currently being researched is the incineration, in a reactor, of the relevant actinides.

The reprocessing of AGR and LWR fuels would, in general, be similar to the reprocessing of Magnox fuels, although, for a given tonnage of thermal-oxide fuels, the radioactive content is about ten times higher. A THORP would therefore separate more plutonium and uranium and would generate considerably larger and more troublesome quantities of waste. The additional production of plutonium would be about 40 tonnes; the volume of highly active waste some 30,000 cubic metres. According to the estimates prepared by BNFL, levels of contamination from a THORP plant operating alongside a refurbished Magnox plant would be reduced to between a half and a quarter of present levels. Overall, they were in 1975 running at about a third of the present maxima laid down by the International Commission on Radiological Protection (ICRP).

If THORP is built and construction goes according to time-table, it will begin operations in the late eighties. The AGRs now operating or under construction (there may also be another two begun in the next five years) will produce some 3,150 tonnes of spent fuel by the mid-nineties. Spent fuel will continue to arise from those AGRs at the rate of 200 tonnes a year, so that, by 2000, if no reprocessing took place, more than 4,000 tonnes would have accumulated, to which would be added spent-fuel arisings from prototype reactors. A total of 6,000 tonnes is a broadly agreed total for the end of the century from UK reactors.

World arisings will be very much larger. Excluding communist countries, the USA, France and the UK, the total estimated for the year 1990 is 20,000 tonnes. The combined French and American arisings might be about the same. The great majority of all these wastes, if THORP is used, might thus be treated in a relatively small section of Western Europe – at Windscale in the UK and at Cap la Hague in western France. The alternatives are to store unreprocessed waste permanently or to confine it to short-term storage until other disposal options have been researched and pursued.

Experience with the reprocessing of thermal-oxide fuels is limited. Between 1969 and 1973, BNFL reprocessed only 100 tonnes of spent fuel from thermal reactors. The plant in which it was reprocessed was shut down (see Chapters 3 and 7) in 1973 after an accident involving a radioactive release and has remained out of service since then. Planned modifications and refurbishment could be carried out by late 1979, providing reprocessing capacity into the middle of the eighties, but there would be little or no margin for breakdown or accident according to the company. Existing plant, in any event, would be incapable of dealing with the spent-fuel arising to which the company is already committed for overseas customers. The application, therefore, is for a plant handling 1,200 tonnes of fuel per year, half of which would be from UK reactors, the other half for contracted and likely arisings from elsewhere – notably Japan, which has asked for a total of 1,600 tonnes to be reprocessed by BNFL.

3 Some Physics, Some Economics

Before the Windscale inquiry began, it was commonly supposed that objections to THORP and to nuclear power stations were, and would remain, rooted in a popular fear of radioactive pollution of the natural environment. Talk of the Windscale site as Britain's 'nuclear dustbin' – a phrase coined by Walter Patterson of Friends of the Earth and used with traditional typographic flair by the *Daily Mirror* – effectively characterized for most people a couple of years ago the attitudes and alarms associated with opposition to the BNFL proposals. But even then, this was simplistic: groups like FoE, with affiliates around the world and dealing with a large number of environmental issues, were evolving a complex of arguments in which the contamination threat was only one.

By the time it came to make its submission on THORP, British Nuclear Fuels Limited was well aware that it would have to do more than answer questions about the environmental integrity of its waste-management systems. A wide range of objections to the application had already been voiced at scores of meetings, in numerous letters to the newspapers, and on programmes transmitted by the radio and television networks. But few of the participants were fully prepared for the weight and width of opposition that was rehearsed in Whitehaven. It is convenient to split the overall case into three parts – the physical and economic; the environmental risks; and the political and the philosophical aspects.

Physically, the opposition arguments can be grouped under three headings: (1) the economic and technical need for an expanded nuclear-power programme; which would predicate (2) the need for and feasibility of the chemical reprocessing of spent fuel from thermal reactors; and (3) the impact its construction and operation would have on local needs and resources.

The first of these centres on the highly vexed question of energy-demand forecasts and the likely roles to be played by different generating systems. Here, it is important to separate two contradictory assumptions. One, held by a majority of those who belong to the antinuclear lobby, is that the demand for energy (particularly in the form of electricity) should and can be substantially reduced without bringing about a deterioration in the quality of people's lives – a contention to be explored in greater detail in Chapter 12. The proponents of nuclear power counter this with *their* assumption that very much higher levels of energy must be available if social and economic stability is not to disappear. Otherwise, they say, the prospect will be one of people fighting and freezing to death in the dark. Nuclear power is seen by one side as misjudgement piled upon mistake. For the other, it represents technological salvation: the only hope of bridging a predicted energy gap.

Until comparatively recently, there was little in the way of documented challenge to this view. Apart from highly specialized reports like those (especially the second) published by the Club of Rome, and Lovins' *World Energy Strategies* – which, though it feels older, dates back only to the end of 1973 – the environmentalist library was a thin one. What argument there was in official quarters tended to revolve round how, when and where the various conventional energy options should be reconciled. Broad agreement on demand-growth rates, predictions based on existing or advanced-prototype technologies and surprise-free suppositions about social trends have been the foundations of government energy policy in most countries.

In five years or so, the situation has changed dramatically, and by the time the Windscale inquiry opened, objectors to THORP were able to draw upon an impressive collection of well-researched scientific dissent to the conventional energy-planning wisdom. Some of the documents produced specifically as evidence to the inquiry are practically standard texts in their own right. At the same time, several submissions made in defence of the BNFL proposals anticipated and catalogued points from the radical analysis of energy futures (one bizarre effect of this was that combatants frequently cited identical references – like the MIT Workshop on Alternative Energy Strategies – in defence of opposing positions).

British Nuclear Fuels Limited claimed throughout the inquiry that a new reprocessing plant was needed to deal with the spent fuel arising from existing thermal stations in Britain as well as that which the company was negotiating to take from overseas – notably from Japan. It chose to reinforce its case by leaning on predictions of a substantially enlarged nuclear-power programme, the upper estimates for which would involve a tenfold increase in the nuclear contribution to Britain's electricity generation. Called on to support BNFL, the Department of Energy put forward a Deputy Secretary, T. P. Jones, who delighted the objectors with one of his opening remarks. Describing the Department's approach to formulating energy policy and the potential requirement for nuclear capacity, he said:

'It is not possible to have a simple, once for all energy-policy blueprint. The uncertainties are too great. We do not know the full extent of world or indigenous fuel resources; we cannot be sure about future costs or availability of supplies, about the rate at which new sources may be developed or how demand for energy may grow ... The Government therefore needs to establish a wide range of energy options and maintain a flexible energy strategy which can be reviewed and adjusted if necessary in the light of subsequent developments.' He went on to outline, however, a rather less than flexible official position in defining a future energy gap and the proposition that it should be filled from nuclear sources.

In the Department's *Energy Policy Review*, published as the inquiry got under way, the long-term-strategy proposals are fairly specific: the two key ones are the commitment to coal (though this is somewhat hedged about with pessimism) and the development of capacity 'to enable a rapid build-up of nuclear power to take place in the 1990s if necessary'. That clearly implies the need for reprocessing plants and other facilities for the preparation and handling of nuclear fuel (including commercial planning and management). It would require some large and irreversible investments to be made. And yet the document, like the Deputy Secretary, concedes that there is a real risk of coming to the wrong decisions.

'The future,' it declares, with no hint of tongue in cheek, 'will always turn out to be different from what we expect', and it goes on to say: '. . . we must recognize the uncertainties and that, if decisions are overtaken by events, we may have to live with the consequences of those decisions, since future decisions cannot set the energy economy on a new course. This requires that our energy strategy should be robust, producing minimum regret whatever course future events take; and that it should not prematurely close options.' Some of the objectors, pointing to the reduction in energy-demand forecasts that has consistently emerged from the past two years' *official* projections, let alone more radical sources, clearly feel that such a closing of options has already taken place.

Indeed, one official witness, Dr John Keith Wright, responsible to the Central Electricity Generating Board for advice on fuel sources and power plant, asserted roundly to the Windscale inquiry that the option on reprocessing 'was closed, like ten years ago' – indicating that, whatever the merits or disadvantages of a THORP, a *nuclear*-energy strategy had been regarded as axiomatic by the most influential of the planners, though they might not represent the considered view of Government, which in its White Paper 'Nuclear Power and the Environment' of May 1977 advanced a qualified commitment. The extent of that commitment, it stated, 'will be resolved in decisions taken progressively over the years in the light of national need and of the acceptability to the country at large of the possible economic, social, and environmental impact of an extended nuclear programme'.

There is a marked difference between that conclusion and those reached by the UK Atomic Energy Authority and the Department of Energy, both of whom say, in effect, that an extended nuclear programme should be developed even if it is later found that this has been a mistake. It may be obvious, but it bears repeating that energy is no different from defence, education or agriculture in the frequently contradictory objectives of those who legislate and those who administrate. One need not read political diaries to know that Westminster can be and often is more responsive than Whitehall. The Civil Service, like the larger industries over which it has controls, is not designed for change: it is an extrapolative machine, not a holistic one.

In practice, this has meant that the only statistical work to which agencies such as the UKAEA and the DoEn will normally attach credence is that adduced on behalf or in support of their assumptions. The Windscale inquiry came as this situation was changing a little, but even with the formation of sub-agencies, like the Energy Technology Support Unit, and the admissibility of work by the Open University's Energy Research Group, there remain yawning gaps in the list of subjects, let alone the data, on which to debate the need for one energy strategy as against another. That fact itself was put forward as part of the case submitted by Gerald Leach, a senior fellow with the International Institute for Environment and Development, who has been carrying out a series of world-energy audits for the last five years.

Crucial information, he told the inquiry, was not available from 'official' sources: 'The numbers of people considering energy scenarios which assume vigorous policies for fuel conservation and renewable resources are pitifully small, as are their supporting facilities, compared to those working on conventional energy-supply technologies and programmes. This asymmetry extends to many other features of energy research and development, including the fact that supply expansion is the aim and business of some very powerful corporations, while energy conservation and renewables have a much weaker and more scattered constituency. This is a main reason why support for and expectations from supply expansion have until recently completely dominated energy forecasting and strategic thinking.'

Leach went on to contrast official figures which, on the one hand, pointed to a steady reduction in predicted overall fuel demands but which, on the other, are nevertheless used to support a case for maintaining or increasing present levels of energy supply. Renewable energy sources – dealt with in detail at the Windscale inquiry by a number of witnesses – are acknowledged in official projections as being capable of providing three times

29

the present *nuclear* contribution to British electricity generation, and yet, as Leach notes, they are excluded from the supply forecasts, where their contribution is described as 'still very uncertain'.

He looked at two other areas where official policy seemed to be based on incomplete or erroneous information. In the first – that of energy usage – the DoEn failed to break down consumption by sectors (such as industry), by efficiency of conversion in turning primary energy into end-use energy, or by the mix of possible fuels. In these, the Department ignored the possibility that the future might be ('and might be made') very different from the past. Secondly, the forecasts relied heavily on two cardinal assumptions, both of which are challenged by conservationists: that energy growth must be closely tied to economic growth; and that future growth trends will be roughly exponential in nature, since they have been in the past.

Since the inquiry, the IIED's work on energy-conservation possibilities has indicated that savings quoted to the hearings might even be on the low side. In Whitehaven, they were greeted with some quiet amazement by BNFL and their supporters – as well they should be if the reformist lobby is to make any progress. In any event, the figures cited by Leach were salutary. Listing a selection of techniques for various energy-using sectors, he claimed that the domestic electrical load for a post-war three-bedroom semi-detached house could be cut by a colossal 86 per cent. For older, solid-walled houses, the savings could be even higher. With 'good housekeeping', industry could show savings of between 5 and 40 per cent. Such a conservation programme, he concluded, would help to solve employment problems (with labour diverted into conservation-related industries) and would eliminate the need for additional nuclear-power-producing capacity.

The arguments were reinforced at great and cogent length by Dr Peter Chapman, director of the Open University's Energy Research Group, who has produced what amounts to a DoEn-in-exile analysis and forecast of energy demand (his *Fuel's Paradise*, also in Penguin Books, is an excellent briefing text and a springboard for some of the arguments he advanced to the

oxide-reprocessing plan
s customers – princip
programme but few if
The Japanese contract
illion, a third of which w
between Japan and BN
arge manufacturing and pl
y pass through and succu
oting the vision of a balar
n theory, might top all otl
try with a distinguished
ess more or less begging to
ce of it, any government we
d such an industry the rigl
d get on with the job – r vide
zed ethical, enviror ntal an
omically benigh governme
the way.
sets of occupationa
BNFL's case – underpir
employment to construct and
any with an annual turnover c
sets of £116 million, employir
an half of whom are so-called 'in
eover, traditionally enjoying re
s with both labour and neighbours,
ng justification for its expansion plar
ze and broad nature of the oppositior
se to lay as much if not greater st
ight appease and even pre-empt the er
A succession of witnesses, of whom Allc
d intellectual leader, insisted through
HORP would be a boon and not a burder
ere concerned. Reprocessing, said Allday,
ecovery and re-use of valuable uraniur
ntained in spent fuel. This represented an eco
e and very substantial contribution towards
n. A single recycle of uranium and plutonium

inquiry). There was, said Chapman, time enough for a more adaptive energy policy in the future: the 'energy gap' forecast was based on unrealistically high demand predictions and pessimistic supply assumptions. The real price of energy would steadily rise (whereas it declined in the period 1950–73), and this would lead to reduced consumption by encouraging energy conservation, improvements in appliance efficiency and improvements in conversion efficiency.

Also to be taken into account were saturation in the consumer market, lower rates of population growth and stabilization of the number of people per dwelling. When oil and gas supplies began to decline, substitute fuels for the lowest-cost heating market could be derived from coal, either as substitute natural gas, or in combined-heat-and-power projects. The least-cost transport fuel, according to Chapman, is likely to be electricity generated from high-energy-density batteries, and he goes some way to show that this could be a cost-effective alternative fairly soon for the passenger-vehicle market. An alternative-energy supply policy, which included wave, solar, coal *and* nuclear components, is, he says, feasible and is likely to be economically and environmentally to be preferred to one based largely on the nuclear option.

The nuclear industry's response to these arguments is not wholly revealed in what was said by way of cross-examination at the Windscale inquiry. There, the reaction – quite properly – was to test in detail the evidence submitted by witnesses like Leach and Chapman. To this end, counsel for BNFL and its supporters closely questioned the analytical arguments entered into, demanding proof of the reliability of this or that energy alternative, its net energy cost-benefit (looking, for instance, at the energy costs of a conservation programme) and so on.

This produced, at least with Chapman, one of the longest and most illuminating exchanges heard on the subject: at once conciliatory and authoritative, he could defend his position without the need for gladiatorial tactics. His position as represented over THORP is, in fact, one of utter independence on the energy demand-and-supply question. Yes – wind generators and mini-Grids scattered about the country could pose environmental problems; so could hundreds of miles of wave machines bobbing up and down the coast. Miners could restrict supplies of coal. People might not take to limited-range battery-powered cars. Large quantities of aluminium would be required for large insulation programmes. New and increasingly popular energy intensive devices might be invented and put on the market. But here were a set of options and a range of possibilities that differed from those offered by BNFL – and why not consider them?

The nuclear industry, and the Department of Energy for that matter, is in a peculiar position here. Some of the critics it regards as idiosyncratic but essentially safe. Professor Peter Odell, for instance, appeared as a witness for the Town and Country Planning Association and, in his capacity as an oil and gas economist, declared that there would not be an energy gap, even without the further development of nuclear and coal supply between 1985 and 2000. Expanded nuclear-power capacity was not required, he concluded, and therefore the proposed reprocessing plant was unnecessary. As he was speaking, arrangements were being completed for him to work for the DoEn as an energy consultant.

There is, perhaps, no paradox here. Scientists and economists, within a broad lee, respect scientists and economists. On where radical social and political change is countenanced those in industry feel annoyed or threatened. They will – or now – argue willingly about the most appropriate energy strategies that ought to be pursued within a generally agreed framework. But they are inhibited from discussing deeper issues that pervade the energy question – issues that were only modestly raised at the Windscale inquiry but which must be faced if the nuclear controversy is to be fully understood. The fact is that a good many of those who forced an inquiry on the THORP application and who will be involved directly or indirectly in forthcoming confrontations do not *want* more energy, whether it is generated in nuclear power plants or not. Amory Lovins, pessimist-in-chief of the energy pundits, synthesized these feelings when he looked at nuclear fusion as a long-term potential source of abundant energy:

'One could argue further that if fusion turned out to be a clean

in thermal reactors would result in the generation of between 30 and 40 per cent more power from the same quantity of original uranium ore. And if plutonium was to be used in FBRs, the energy savings would be fifty to sixty times greater.

Not only would reprocessing and recycling relieve the pressure on other fuels in the remaining years of this century; it would conserve stocks of uranium and therefore reduce the country's vulnerability to increased uranium prices or outright scarcity should resource-diplomacy of the kind associated with OPEC become a feature of the nuclear market. It would also – and this has been a crucial part of the industry's reasoning – conduce to a nuclear-waste-management policy officially considered to pose fewer environmental difficulties. Instead of storing *un*reprocessed wastes, as has been done overseas (and to a certain extent in Britain too) – a policy advocated now by the USA, Canada and critics of THORP at the inquiry – it is proposed that the higher-level radioactive wastes, once separated from other fission by-products, should be vitrified, or sealed into glass blocks, and deposited in 'permanent' storage caverns, to be mined in deep geological strata.

If environmentalists really wanted to reduce energy consumption and avert the risks of radioactive contamination, then THORP, the argument went, was for them. So sure of these grounds were the legal and technical advisers to BNFL that they had not deemed it necessary to provide – or, in some cases, determine whether there were available – data on the economics of *not* reprocessing spent fuel, on the estimates of uranium price and supply futures, and on the physico-chemical integrity of stored unreprocessed fuel.

Clearly *something* has to be done with the spent fuel from existing reactors, which have already produced several thousands of tonnes. But no serious modern study of the problem had been conducted by the industry with a non-reprocessing option in mind. It had not been considered. The objectors, however, wanted it to be, and for several reasons. They wanted proof that the radiological hazards to be associated with a reprocessing/recycle-and-storage route were lower than those resulting from the simple storage – in water or in gas-cooled vaults – of canistered spent fuel more or less as it arrives from a reactor. If it *could* be safely stored in this form until a truly permanent disposal solution might be devised, was this not to be preferred to a route that involves the unproved technology of vitrification (the HARVEST process), the uncertainties of geologic storage and the commitment to a swelling inventory of plutonium?

Only towards the end of the Windscale inquiry was any substantive information made available on the question. Two studies, one done by the UKAEA and the other by the Battelle Pacific Northwest Laboratories, came to the tentative and independent conclusion that unreprocessed fuels could be satisfactorily stored for periods of up to twenty years without serious oxidation or embrittlement of the fuel cans. Fuel from later types of thermal reactors, such as the AGRs, could be kept safely for at least twelve years. The UKAEA study – run at the inquiry's request over a two-month period in 1977 – implied that fuel elements from AGRs could possibly be stored for fifteen years with suitable pond-water treatment and redesigned fuel cladding.

The inquiry Inspector, Justice Parker, was moved to note that it was remarkable that this was the first piece of research that had been done on the integrity of pond storage. Until then, he assumed, '... they just had a look at them [stored cans] every now and then and said: "They seem to be all right – there seem to be no problems." Might it not be a good idea if somebody *did* something?' Counsel for the Friends of the Earth, Raymond Kidwell, QC, answered with some force and relish – for the issue had taken up many many days of evidence and cross-examination: that, he said, 'is what we have been pleading for' – and he went on to make a further point. FoE was not opposed unreservedly to the storage, in granite mountains or beneath the ocean floor, of spent fuel. But if the geological and engineering problems of doing that were solvable for glass blocks, they were equally so for canned spent fuel. The worries that remained were of environmental discharge and the accumulation of plutonium, one of which was not solved by reprocessing, the other of which was exacerbated.

FoE has argued strongly that, in the short-term – twenty-five to thirty years – there is no technical necessity to reprocess spent fuel and that safe and suitable alternatives exist, one being that of bottling individual irradiated fuel elements and storing them in isolation. If reprocessing *was* a vital need, say FoE, existing plant at Windscale could be used to handle UK fuels and a decision on the construction of new facilities easily deferred for ten years (by which time dramatic technical, political and energy-demand changes could well have taken place). Even if the British Government decided to approve a commercial fast-breeder reactor (CFR-1), present and potential stocks of plutonium would be available to fuel it. If a full-scale FBR programme was sanctioned, construction of the accompanying reprocessing plant need not begin until the stations themselves were ordered, and that, by common agreement, is not likely to be before 1990.

BNFL is in something of a bind over the question of technical satisfactoriness, and evidence submitted to the inquiry was not wholly helpful. A major part of its existing reprocessing facility – the so-called Head End plant – went abruptly and alarmingly out of service in September 1973, when there was an accident involving the inadvertent attempt to fill a vessel already containing highly active residues. In all, BNFL has reprocessed only about 100 tonnes of spent uranium oxide fuel – a fifth of the total quantity reprocessed in the world (excluding China and the USSR). Yet the THORP application is for a facility that would handle at least 1,200 tonnes *per year*. The objectors have, cruelly but inevitably, suggested that – if the Head End plant can be refurbished – a ten-year abeyance would enable the company to demonstrate that it has a mature and reliable technology from which to develop the much larger project.

There is conflicting evidence on the point. In principle, according to the Nuclear Installations Inspectorate, the Head End plant could be operated for a further period of between ten and twenty years, handling a throughput of 400 tonnes of spent fuel – figures that lend some strength to the objectors' suggestion. But BNFL, though uncomfortable at the prospect of revealing the precise nature of its difficulties, insists that restoration of the Head End plant is not a sound or economic proposition. Word from inside the Windscale works indicates that this is so – leaving BNFL unable to prove to the opposition's satisfaction that the massive scaling-up operation represented by THORP would be a trouble-free operation.

When confronted with past failures, BNFL – like other advanced-technology companies – has a stock response, which is: 'Yes – things went wrong, but we were provided with invaluable operating experience, and similar mistakes will not be made in the future.' Thus, Con Allday declared at an early stage in the inquiry that he was 'completely confident' that the company had the competence to engineer a THORP successfully and safely. Having already indicated that he was unimpressed by North American trends towards the storage of unreprocessed waste and the development of techniques for such storage, he was equally less than convinced that reprocessing in the USA had been held up by serious technical problems.

Allday believed that the delays associated with the Barnwell plant in South Carolina were mainly the result of economic and regulatory constraints. Even if this were wholly true, it would not much help BNFL's case: the cost overrun on Barnwell, when the plant was denied further funds, was running at 1,000 per cent. If the Americans could make that kind of economic blunder, why should the British be able to do much better? Allday told the inquiry that there was not, in his judgement, a high risk of a major cost overrun. THORP would, in any case, be designed with the help of a lot of information based on American and French experience. But the Americans *had* run into technical difficulties as well as financial ones – difficulties described as 'serious' in a pro-nuclear journal. That, Allday concluded, was because of the deficiencies of a particular process used in the USA and not elsewhere.

What about experience with the advanced gas-cooled reactors, he was asked: surely those in the nuclear industry who were responsible for designing, planning, implementing and building the AGRs had equal confidence in their skills ... and they had been subject to disastrous cost and construction problems? 'No,' said Allday in reply to Kidwell's questions: 'I don't accept that. Dungeness B I think was a disaster' (it was, in fact, one of the biggest engineering fiascos in power-station history) '. . . but I think the builders of Hinkley and Hunterston did a remarkably good job. I think the nation will eventually be very grateful to the builders of those two stations.'

Three months after Con Allday said this, the Hunterston B reactor was shut down after an 'unprecedented failure' in which 1,000 gallons of seawater leaked into the reactor. The total possible cost of the accident was put – in December 1977 – at £14 million, which would have the effect of raising electricity tariffs for consumers by 2 per cent. Generating Board officials were at pains to stress that the fault owed nothing to an inherent design weakness in AGRs. But it is almost axiomatic that if a plant is large and complex, then so will be the costs and consequences of a failure – however 'low-brow' in character it might be. At Hinkley Point, a similarly run-of-the mill fault caused the power plant to be shut down from July to December 1977. A pipe supplying make-up feedwater to the main coolant system fractured. Hoses had to be rigged up to play cooling water on the concrete shielding of the primary reactor. Again, the failure itself was a simple one; the impact was not.

The point was underlined – some might say laboured – by other witnesses at the Windscale inquiry, who catalogued accidents and failures in the nuclear and other industries in which the most common causes were procedural or logistical rather than operational and in which design had not allowed for operator error, breakdowns of a simple mechanical or electrical kind, and circumstantial factors beyond the scope of conventional engineering-design practice and philosophy. (A favourite, though inconclusive, example is that of the destruction of the Flixborough chemical plant, where leaking cyclohexane gas is thought to have ignited after having been drawn into the cylinders of an idling diesel engine in a parked truck: prohibitions on naked lights, ferrous tools and electrically operated appliances were not enough.)

However good or poor an illustration that is, the nuclear industry cannot pretend to be immune from the good old-fashioned law which says that if things *can* go wrong, they will. The critics' case is that the consequences of failure must be minimized by going for relatively small and simple technologies, and in this they are no more strongly opposed to the Windscale expansion plan than they are to projected petrochemical complexes, liquefied-natural-gas terminals and so on. Much time was spent at the inquiry in discussing risk assessment, but a recommendation for or against THORP could not, in all conscience, be made on the basis of what was said. The one-in-a-million chance of catastrophic accident is a rather meaningless coefficient, though it was much quoted (together with the risk-equivalence of 10^{-6} for travelling by road for fifty miles, rock-climbing for $1\frac{1}{2}$ minutes, or simply living for a further twenty minutes after the age of sixty).

An aircraft loaded with high explosive *could* crash into a nuclear plant when all the workforce were on strike and when all supplies of cooling water were cut off – but it is profitless to attempt to express the probabilities statistically. Dr Brian Wynne, who appeared at the inquiry for a group known as Network for Nuclear Concern and whose learned-on-the-spot abilities for cross-examination drew praise from the legal professionals, rejected the validity of arithmetical comparisons. He argued simply that the risks imposed in a relatively involuntary way by agencies of a centralized kind should be seen to be no greater than those accepted from choice – such as those involved in travelling by car.

The question of risk distracted attention from dispute over the economic case for building THORP, a matter to which the combatants returned again and again as one set of figures were produced to challenge another. Since we are looking at issues, rather than at the fine details raised, there is little to be gained from following – even in synopsis – the great expanse of ground covered by the adversaries. The argument falls into two halves. The first concerns the economics of nuclear power itself and the component part played by reprocessing. Though the figures change month by month, the relativities as cited by each side remain roughly the same. Thus BNFL, in its annual report for 1976–7, quoted DoEn statistics as showing that comparative generating costs for the principal fuels were: oil, 0·8p per kilowatt-hour; coal, 0·74p/kWh; nuclear, 0·48p/kWh. The first

AGR stations, said the report, would join earlier nuclear power plants as producers of 'the cheapest electricity available'.

These comparisons are strenuously denied by many outside the nuclear industry, and indeed by some who are not necessarily opposed to the building of THORP. It is widely felt that there are large gaps in the cost-analyses prepared and published by and on behalf of the nuclear industry. At the least, there are ambiguities. Compared with coal and oil, nuclear fuel itself is cheap, but the costs of using it are high and are commonly masked by the way in which the financial accounts are presented. The true cost, for instance, should reflect the very considerable levels of public expenditure devoted to the research and development of nuclear power.

These funds are nominated in such a way as to conceal or minimize the direct debit to nuclear power. As well as providing substantial capital grants and loans to the nuclear industry – including a £1,000 million subvention made accessible to BNFL in 1977 for its plans – the Government provides a large proportion of the research and development costs. In 1976, £95 million were available for nuclear fission R&D, compared with £6 million on fusion and a total of £7·5 million on all renewable-energy projects. Taking the state and private sectors together, the proportion of national R&D in Britain devoted to nuclear power is 65 per cent. These figures are only partially used in the official calculations of unit cost, and there are others whose absence or incompleteness blur the reckoning.

Among them are the costs, yet to be faced, of decommissioning nuclear plants at the end of their useful life (see Postscript) and providing virtually permanent radiological safeguards (measured in energy inputs alone, some critics have argued, these costs could exceed the relatively short-term benefits to be gained from operating the plants). Heavy associated costs for electrical-energy distribution systems suitable for nuclear power are not effectively included; nor are the subsidies which the industry makes to the defence sector through the provision of weapon-grade products. Other 'hidden' costs listed by the antinuclear lobby are those for building overcapacity into nuclear generation (to allow for safety shutdowns and slowdowns), the administration of nuclear regulatory agencies and organizations, the high prospective costs of mining and refining lower-grade uranium ores, and inadequate amortization accounting for existing plant.

According to Colin Sweet, a senior lecturer at South Bank Polytechnic in London, the official unit costs quoted by BNFL and the CEGB for nuclear power are between three and four times too low. In an extensive analysis of costs presented to the Windscale inquiry, he suggested that the true price of AGR-generated electricity would not be less than 1·5p/kWh – compared with BNFL's figure of 0·48 – if the costs were properly broken down. Later in the inquiry, after correspondence and discussion with the company and the Generating Board, he revised his estimate, upwards, to 1·83p/kWh. A similarly large discrepancy was offered for the influence of reprocessing on final electricity costs. BNFL put it at around 8 per cent. Sweet and others point to figures three times higher.

Even Amory Lovins, whose energy-cost analyses are probably the most exhaustive and careful, treats reprocessing costs as one of the significant but barely quantifiable components in the overall delivered price. It is one of 'some fifty-odd main variables, of which few are known to within a factor of two, all are disputed, and none is orthogonal [independent of the others]'.

Not surprisingly, the economic argument over THORP itself became an impenetrable, irresolvable one. It involves fiscal dynamics which not even the large energy industries have control over, still less complete understanding of. Neither do the objectors – but they did not take it upon themselves to put forward a case for expanding the nuclear-power programme or for reprocessing and clothe the case in a selection of statistically favourable data. It was left to the objectors to produce for themselves, winkle out or make reasonable guesses about the bases on which profitability could be calculated. By the end of the inquiry and certainly by the spring of 1978 when FoE published a fresh set of costings for the proposed plant, more economic data were available from the objectors than from BNFL and its supporters.

Those data indicate that THORP, at best, will be constructed on a combination of optimistic economic factors and forecasts, in which uranium prices, storage costs, reprocessing overheads, discounting levels and periods, exchange rates and capital-cost increases together conduce to a net *national* gain over the ten-year-minimum operating life of THORP. Not to reprocess at all would undoubtedly save Britain, in isolation, a lot of money – somewhere between £300 and £400 million. But the company insists that *un*reprocessed wastes would eventually have to be vitrified, in which case the costs would be even greater than those of reprocessing.

Two other factors add to the complexity. If plutonium recovered in THORP is not used as a fuel in FBRs, were they to be built, then a significant economic advantage is lost. But no one denies that the company will make at least *some* commercial sense out of its contract for the reprocessing of Japanese spent fuel, whatever else is likely to alter. The terms of that contract are almost those of a gift, in which the Japanese electric-power utilities bear all the technological and operational risks, and where repayment liability is incurred only if BNFL fails to build and operate THORP with the capital – £140 million or so – specifically advanced by the customer for the purpose. Cogema, France's counterpart to BNFL, has signed a similar contract for reprocessing half of the projected Japanese spent fuel at Cap la Hague.

The arguments are somewhat akin to those that have attended the Concorde project – though the nuclear industry can and does protest that there is a world of difference: the Anglo-French SST has been a financial failure for want of customers, they say, whereas reprocessors are faced with all the business they can handle if governments and lobbyists will only let them get on with the job. But there is an analogy in that national, still less global, interest is not the same as sectoral commercial interest. Prosperity for BNFL, its employees and its shareholder the AEA might be attainable only at the expense of other parts of the economy and would distort the future energy scene by diverting R&D resources on a large scale and continuously to this one technology.

As part of the economic case for building and operating THORP, the applicants suggested that as many as 1,400 additional jobs would be created by the new reprocessing plant

and other permitted developments on the Windscale site. About
1,000 jobs were to be associated with the THORP project, and
they would, said the company secretary Arthur Scott, be offered
wherever possible to local people. He thought that roughly half
the 'extra permanent jobs' would be taken by people already
resident in the area, in which unemployment runs particularly
high. The West Cumbrian coast has seen a sharp decline in almost all its traditional industries – fishing, farming, coalmining,
steelmaking and shipbuilding – and in parts of the region young-
male unemployment is as high as 25 per cent. On the face of it,
the prospect held out by BNFL is a very good one.

The local authority, however, indicated that *its* own estimate
of employment potential was somewhat lower. Over a ten-year
period, the proposed plant would be likely to generate no more
than 700 new jobs, of which only a third would be drawn from
the local unemployment pool and by school and university
leavers. Objectors suggested that BNFL has not, in any case,
gone out of its way to recruit from that pool in recent times.
During the course of the inquiry, for instance, the company was
advertising in the London *Sunday Times* for a canteen manageress
– a specialized and responsible position, to be sure, but not one
that should have been all that difficult to fill locally.

Cumbria County Council's Structure Plan saw the problem in
this way: 'Should we put employment first always, or should we
balance this against the Cumbrian environment and pollution
risks, giving each a weighting?' Fraught with wider implications,
this is the kind of question over which local government is
agonizing throughout the industrialized and developing world
as the orthodox bases for employment crumble. What they want
– and it is expressed universally – is diversity; what they are
offered, with the sticks and carrots of federal intervention, is
concentration. The one-company town or single-industry
region, with its wide environmental and economic implications,
seems to many to be even more of a mistake where – as in the
nuclear industry – installations have a fairly short predicted life.

According to the Director of the Town and Country Planning
Association, David Hall, the risk of even higher unemployment
would be considerable: '... the public investment that would

have gone into the services and housing to support an albeit temporary workforce might be wasted, and the opportunity would have been delayed even further to introduce a more diversified and permanent industrial base'. This is an argument that might have been developed at greater length, perhaps, for it is central to the debate on modern social and industrial policy planning and the problem of growing structural unemployment in technologically advanced countries. In the end, though, the issue over jobs at Windscale rested on two points.

The first is that the effective subsidy for 1,000 jobs, assuming that was to be the number specifically required for the project, appears to be about £70 million – or £70,000 per job from public funds. Employment subsidies in areas of need are commonly less than £1,000 per job. Even the Ford Motor Company's development in South Wales, which has come under criticism as taking an injudicial allocation of national economic aid, calls for around £14,000 assistance for each job created – the highest large-scale subsidy ever made. The second point follows: that employment is a thin-ice argument on which to skate in support of the Windscale proposals. By the end of the inquiry, BNFL had effectively dropped it – as had the Inspector, who concluded that '. . . it does not seem to me that the amount of extra employment is such that it should override any degree of risk to the public which should properly be regarded, employment apart, as an acceptable risk'.

4 It's Safe: We Say So

'There is no common knowledge between the inexpert members of the public who are worried, and the experts who interpret the hazards. There is no untutored commonsense image of a hazard, from which ordinary people can derive some feeling of control over it, and the consequent sense of security. Hence, technological hazards once discovered have an especially dread quality ... those protesting may well develop an intensity of emotion that seems misguided or irrational to those in positions of safety, who can approach the problem in a detached scientific manner. Those citizens and their advisers who have trained themselves in the technicalities of the hazard are liable to be scorned by the qualified experts in spite of possessing full competence on the problems in question.'

That passage, from a 1976 Council for Science and Society publication *The Acceptability of Risks*, is an uncannily apposite one for an inquiry that, a year later, would be dealing with the containment of radioactive substances. BNFL came to the Windscale hearings prepared to defend its record and its outlook on several distinct aspects of environmental health. First it was concerned to allay any fears among its workers about their occupational wellbeing. Then it wanted to reassure the local general public, and the interested lay public elsewhere, that they and their natural environment were in no danger of contamination from the transportation, processing or disposal of radioactive materials. Lastly the company knew that it had to face opposition from those 'who have trained themselves in the technicalities of the hazard'.

In one of the earliest submissions, made by Peter Mummery, BNFL's director of health and safety, a paradox is revealed. In his review of radiological-protection methods and standards, he

46

said: 'The development of nuclear weapons and the subsequent development of the nuclear power industry demanded an increased attention to the known risks from radiation and to the controls and limits required to provide for the health and safety of those engaged in the industry and the general public ... much work has been done over the past thirty years or so, and this work is continuing on a worldwide basis. However, general experience has confirmed that the foundation of safety controls was firm. One indication of this is that the number of deaths of those engaged in the nuclear power industry which are known to be attributable to radiation is very small. The nuclear industry has not followed the pattern of learning from bitter experience involving a history of injuries and premature deaths due to exposure to toxic materials, and then of applying standards and instituting controls.'

So far as it goes, that is a sincere and truthful statement. It is futile to attempt to prove or to imply that the nuclear industry is or has been an occupational disaster area or a cardinal polluter of the environment. Occupationally and environmentally, coal has probably been the worst culprit – the cause of deaths and maiming in the mines, of gross aerial pollution and consequent respiratory and pulmonary disease from coal-burning in homes and in industry, and of vicious aquatic and soil contamination from the coal-tar chemical and dye industries. Other industries have bad, if not quite so shocking, records. And if an accusing finger is to be pointed at a single technological development, it will pick out the motor car, which – with the possible exception of gunpowder – has been the cause of more deaths, injuries and disease than has any invention in the history of the human race.

By contrast, the numbers of those likely to be killed or measurably harmed as a result of past or present activities in the nuclear reprocessing industry are very low. Nearly a quarter of a million people have died in car accidents *in Britain since the last war*. Hundreds of thousands more have been seriously injured. The costs of physical and human wreckage on this scale are beyond practical calculation. In addition, the car has imposed fearful environmental burdens on society, filling the cities' air with noise, nitrous oxides, unburned hydrocarbons and lead. A profligate consumer of space, material resources and human skills, the car has visited far more specific sins upon us than could ever be levelled at the technology of nuclear fission.

But the social ecology of which the car is part is also that which underpins assumptions concerning the need for and the safety of an expanding nuclear-power industry. The controls and safeguards are established within a nanny-knows-best approach both to the framing of legislation and making of recommendations, and to the design and use of specific protection mechanisms. Industries as large as those involved with cars or nuclear materials play a semi-governmental role in the setting of operational and product standards: large subsections of the state machine – regulatory agencies, proving laboratories, research institutions, monitoring units and so on – effectively become a part of the industry they are set up to serve. There is nothing conspiratorial or sinister about this: it is a thoroughly modern organic convention.

The risks and potential deficiencies of such a set-up are obvious, however, and the most important is that no institution (excluding rather special ones like the Family Planning Association) is going to determine for itself operational thresholds so high that it runs the risk of going out of business. A final analogue from the car industry may be pertinent. The motor manufacturers and their infrastructural agencies well know that the combination of car, driver and open road is ergonomically and cybernetically unsound: a series of accidents looking for somewhere to happen. But the only real solution – a complete switch to public transport – is commercially unthinkable and, politically, beyond their grasp. So the paradox in what was said at the inquiry about nuclear health and safety is this: the industry, recognizing that it embraces the innate potential for catastrophe, cites that recognition as *proof* of its ability to avert or avoid such catastrophe.

Ten major submissions, for the objectors, were made to the Windscale inquiry on the effects of radiation on the environment, the general public and workers in nuclear installations. Most of these either began or were eventually bogged down in highly specialized and – it often seemed – pernickety dispute over figures. The length of this chapter, in fact, does not mirror the time devoted at Whitehaven to discussing the environmental integrity of existing and proposed BNFL facilities. A number of witnesses, or the groups for whom they appeared, decided that the most effective way to put their arguments against THORP was to produce detailed rebuttals of official data on radioactive discharges, the incidence of radiation-induced cancers and the nature of radiobiological pathways; or, alternatively, to introduce data of their own which suggested ignorance or concealment on the industry's part.

What is most at issue is the extent to which so-called 'routine discharges' of low-level radioactive waste are likely to disrupt the natural ecology and/or interfere with human health after being passed as effluent into the air or water. They include a large variety of radionuclides – based, among others, on iodine, krypton, ruthenium, caesium, tritium and strontium, to which, after radioactive decay, other elements from more highly active sources, such as plutonium, americium and curium may be added. These are allowed to be discharged from plants such as Windscale within prescribed or 'authorized' limits. Opponents argue that too little is known about the impact of these substances on the environment to permit any increase in volume. Some caution that, while there may be no presently proven risk to man himself, the contamination and subsequent possible reconcentration of radioactive pollutants by other organisms, especially where resuspended plutonium is concerned, could be equally hazardous. A large number of objectors, whatever their individual stance, are unhappy about the way in which standards are set and met.

To some extent, their case *must* have some substance. They do not, for the most part, claim that the release of radiopathogens to the environment is a unique hazard or that the industry is singularly neglectful of its responsibilities to prevent that hazard. What they do say is that the nuclear industry is no better able than any other to guarantee conformity with the standards of protection laid down by international control bodies, and so should not be allowed to go ahead with expansion on the scale proposed, for instance, by BNFL. The Royal Commission on

Environmental Pollution in its Sixth (Flowers) Report observed that, though further research work might require the amendment of radiological protection standards, '. . . there does appear to be more agreement over what can be tolerated than there is for most other pollutants'.

No one, I believe, will deny that, but there remain many who contend that agreement on tolerance levels is a vitally different matter from that of setting, authorizing and monitoring truly satisfactory standards and procedures for keeping within those levels. The argument is about human and organizational fallibility. It is about the continual pressures – preponderantly innocent ones – to treat environmental and health hazards as secondary matters. This risk is palpable: a company like BNFL is, after all, not in business to research and develop safety, hoping to reprocess a little spent fuel on the side. Equally, the institutions that *are* charged with establishing and vetting radiological safety are part of a network in which the functions of law-maker, policeman, judge, jury, appeal court and all may be performed, effectively, by the same individuals or groups of individuals.

It is, largely, a question of credibility. One of the most eminent expert witnesses who appeared before the Windscale inquiry was Dr Vaughan T. Bowen, an American geochemist now working as senior scientist at the Woods Hole Oceanographic Institution in Massachusetts and one of the world's foremost authorities on aquatic pollution. He is not opposed to nuclear power: indeed, he worked on the Manhattan A-bomb project and was associated with the choice of Windscale as a site for the first commercial nuclear power plant – Calder Hall.

In his evidence for the Isle of Man Local Government Board, however, he lambasted some of the bodies responsible for monitoring and controlling radioactive discharges, and singled out the Fisheries Radiobiological Laboratory at Lowestoft (a branch of the Ministry of Agriculture, Fisheries and Food) for criticism. The FRL, whose brief is to ensure that the marine contamination from such discharges is well below danger thresholds, was – said Bowen – 'defending Windscale and related operations, rather than examining them with care': in its reports, the Laboratory had consistently failed to ask " 'why, where or how?'". What was needed was monitoring directed towards evaluation of future risks, 'rather than, as at present, simply towards establishing the lack of harm produced in the previous twelve months . . .'

Comment of this sort is not lightly received by the industry or the institutions involved, and a strenuous refutation was made to much of what Bowen had said. In fact he was only stating from an authoritative scientific standpoint what many environmentalists are saying in more general and philosophical terms. His challenging of the methodology of data collection and analysis leads naturally to a second criticism in his submission: that the data are commonly kept within specialist Government and industry confines. He went on to express concern over the practice 'of subjecting the public, or a fraction thereof, to health hazards about which they are left ignorant, and to which their response is not even being studied . . . the public reaction to the information brought out should be taken into account'.

Justice Parker rejected any inference from Bowen's detailed evidence that there might be significantly higher levels of radioactive contamination in the Irish Sea than had been supposed. But he did take note of remarks about informing the public – particularly this one: 'Unless our attention is directed not merely to maintaining safe operating practices, but to convincing most of the public that we *know* these are safe and helping them understand how we know these are safe, then there is a very real possibility of democratic refusal of further development of our nuclear potential' (emphasis added). In his report, Parker declared that this observation 'carries weight', and went on to suggest possible changes in reporting procedure for the International Commission on Radiological Protection.

I am not sure that Parker fully appreciated what Bowen was asking for, however, since he appears to have ignored or rejected no less explicit but more radically expressed pleas by others – notably Dr Brian Wynne. Wynne, again not taking an exclusively antinuclear approach, also made it plain that he was not asserting that levels of environmental contamination from Windscale were dangerously high. They might or might not be so. His main thesis was that the regulatory processes and institutional arrangements on which the processes are based 'give insufficient grounds for confidence that proper control is being maintained and will be maintained in the future'.

Among other aspects of the problems, Wynne looked at the cost-benefit formulas that are applied to risk. They involve, broadly, the trade-off of (1) the estimated social and health costs of a given environmental discharge, set against (2) the costs of installed control technology needed to avoid that discharge. Wynne notes that, in theory, when (1) exceeds (2) the necessary technical measures should be taken, but that, though elegant in principle, in practice this is a complex approach because of the meaningfulness of 'social' costs. The difficulty is overcome by reformulating (1) as 'the price society should be willing to pay to avoid a given discharge'. That, of course, is still quantifiable only in the most approximate fashion, and so, in the words of an FRL witness, 'one has to rely on professional judgement'.

That argument, says Wynne, is used to justify the present 'private' system of setting discharge limits. He goes on to say that, without an explicit, systematic and public procedure for setting discharge limits, 'there is no guarantee that BNFL will not be able, for purely commercial reasons, to reduce the costs of reprocessing by negotiating [with the standards institutions] unacceptably high discharges'. This is not an argument about corruption any more than it is an attack on the scientific rigour of work carried out by individual environmental protection agencies. It is an appeal to break down the isolation in which radiological-protection standards are set and to initiate independent and complementary research which will 'serve the role, vital to science, of creating a properly pluralistic distribution of investigation and interpretation'.

In spite of a very considerable volume of evidence from BNFL and the regulatory agencies which do monitor environmental discharges from nuclear plants – evidence suggesting that adequate safeguards *are* in force and that contamination of the general environment has not reached worrying levels – fears are far from allayed. At the Windscale inquiry, these fears were based more on suspicion than on hard fact but they were real and forcible none the less. Dr Robert Edmund Blackith, a fellow of

Trinity College Dublin and a lecturer in the zoology department there, was called by the Windscale Appeal Group to speak on behalf of the Irish Conservation Society and the Wexford Nuclear Safety Association – which he did, while emphasizing that his remarks were not confined to Irish interest.

Ireland, he observed, is one of many countries which run an environmental risk from routine discharges occurring in the nuclear industry but which do not benefit from any advantage there might prove to be from nuclear power: '. . . face to face with Windscale, [we] receive liquid discharges with no option to deflect them'. A large-scale release of gaseous or particulate effluent, following a serious accident at the plant, said Blackith, could spread radiation damage over a large part of Ireland.

The nuclear lobby would no doubt say that such risks were negligible. 'So, until recently, did the asbestos lobby, the cigarette lobby, the food-additives lobby and the pharmaceutical lobby.' In most such instances, he explained, there was a long latent period between exposure to contaminants and toxins and the manifestation of diseases that resulted – particularly cancers. That latent period, which might amount to three or four decades, made the establishment of a legally satisfactory association of cause and effect difficult. So did any interaction or synergism between two or more carcinogenic factors (an argument stirred at rather fruitless length in other evidence): cigarette-smoking, for instance, might greatly increase susceptibility to the effects of asbestos or radiation on human lungs.

He referred to American studies which appeared to show that the levels of radiation to which the general public were exposed and which arose from the nuclear-fuel-processing industry were known to be causing an increase in the incidence of cancers, of leukaemias and of genetic damage – although these increases were ' "frequently not demonstrable" '. He took the polluter-must-pay principle a stage farther by suggesting that 'murder by technology' be made a criminal offence – and quoted precedents abroad for this – a demand that had been echoed by a senior Irish trade-union leader, John Carroll, in a speech to the EEC Social and Economic Committee when it was examining a proposed European Nuclear Safety Code.

In France, he noted, some steps in this direction had already been taken. In March 1977, the former director of the Centre d'Études Nucléaires at Grenoble, together with his Chef de Service de Protection de l'Environnement, were sent for trial after a detailed technical report had been received by the examining magistrate in a case concerning pollution of sub-city waters by the element antimony. If the principle of personal responsibility were expanded and extended to all countries, Blackith concluded, 'more care might be taken before confident assertions are made that there will be no harm to the public'. By implication, industry in general, and the nuclear industry in particular, could be forced to abandon its belief that ' "The solution to pollution is dilution" '.

Blackith's suggestion is a superficially attractive one, but its practical prospects must be slight. The legal profession and the courts would be swamped if legislation on this basis were introduced, and the difficulties of apportioning blame and responsibility would be immense and painful. A more realistic approach is probably that advanced by Brian Wynne, whereby the organizations responsible for safeguards are restructured to ensure genuine independence and public involvement. But Blackith's submission is important in that it reflects the anger and resentment that popularly surrounds the question of pollution – especially a form of pollution to which the human senses cannot make timely and quantitative response.

That resentment is unlikely to be diminished much by the reassurances given during the inquiry and synthesized in the Parker report. When figures were produced revealing substantial and abrupt increases in radiological contamination (for example, a thirty-nine-fold increase in the release of iodine$_{131}$ over three months in 1972, and a twelve-fold increase in the discharge of caesium$_{137}$ between 1970 and 1975), the official response seems to be: 'Ah, yes – but there were special circumstances obtaining at that time. In any event, the levels of contamination were still within the limits laid down by the ICRP.' Parker says that such examples demonstrate that 'there is a large margin of safety and that even when serious errors are made the results need not endanger the public'. It is hard to accept the reasoning here. If a twelve-fold or thirty-nine-fold excess is possible, there is no technological reason why it might not be 120-fold or 390-fold over some other period or on some other occasion. That other occasion, objectors argue, is made far more likely by the construction of a plant at least ten times the size of existing facilities at Windscale.

The question of occupational risk is perhaps the least tractable of all and raises many issues to which the Windscale inquiry could not really address itself but which are part of a general argument about working procedures in high-technology industries. Thus, employees may take risks the nature and severity of which they are not aware; they may be instructed only partially on the context within which they perform their duties; supervisory inadequacies and disputes with management may lead them to ignore or even thwart safety systems; instrumentation may develop unrecognized faults or be inherently open to misunderstanding; operator faults may go unreported or be concealed for fear of disciplinary action. A permutation of these and other factors could put employees and the public in jeopardy.

Only one study seems to have been conducted into the question of safety as it affects employment rights and conditions in the nuclear industry. This is by Roy Lewis, a lecturer in industrial relations at the London School of Economics, and was published in the March 1978 issue of the *Industrial Law Journal*. Relying in part on material gathered by the Socialist Environment and Resources Association (SERA) at Windscale – used in unchallenged evidence at the inquiry – Lewis notes that what, in other industries, would be called 'danger money' is called an 'abnormal conditions allowance' in the British nuclear industry, payable to certain workers who are liable to require decontamination from exposure to radiation. Other special payments include an 'irksome' duty allowance to compensate for the wearing of protective clothing; and a lump-sum payment for workers who have to be taken permanently off shifts involving work with plutonium because they have reached maximum permitted levels of exposure.

Anxiety over safety, says Lewis, expresses itself mainly in monetary demands 'perhaps because of the difficulties of eliminating the risks . . . certainly the pay structure reflected the inherent dangers of nuclear employment'. At the time of the inquiry, BNFL unions were negotiating claims for damages for the families of two plutonium workers who had died, it was alleged, as a result of their contact with the substance. Shortly afterwards, an award of £22,000 was made to the widow of one of the men and £8,000 to the second, though the company denied any liability in the latter case. Shop stewards at Windscale have long considered that all radiation-related sickness should be classified as an industrial disease for the purposes of compensation, and their argument has led to talks on an automatic-compensation plan, similar to that operating in the coalmining industry.

Workers in the industry are, understandably, reluctant to discuss the subject with outsiders: their loyalties and vulnerabilities are an obvious barrier to openness. But there is little doubt that many employees do harbour fears for their safety. One instance quoted by Lewis is the widespread belief among reprocessing workers that fuel rods are allowed to remain in some reactors for six months longer than the design specifications indicate and that the cladding, when removed for reprocessing, is unacceptably fragile and excessively irradiated. There is also a feeling that health records have not always been adequately kept and monitored (a point dealt with at the inquiry by BNFL and by the epidemiologist Dr Alice Stewart).

Management is universally suspected of secretive behaviour, and this usually exacerbates mistrust. In the case of the nuclear industry, the management may not invariably be in a position to help itself. Regulations due to come into force in October 1978 imply that national-security requirements will override the free and open consultation on safety questions which unions have pressed for. These regulations (contained in Section 2 of the Health and Safety at Work Act) specifically excuse the employer from the duty to give safety representatives (of recognized unions) 'any information the disclosure of which would be against the interests of national security'. That might cover, for example, information about the manufacture, storage and transportation of plutonium or enriched uranium. SERA's view is that, at a time when employee participation in safety is the policy of Parliament and industry, 'the proposed development at Windscale will create a major new area where public policy is likely to be frustrated'.

Even without problems of this kind, workers in the nuclear industry are beset by managerial assumptions, just as common elsewhere, about their need to know the *precise* nature of their exposure to calculated risk or – some observers might add – the nature of their role in the whole operation. An us-and-them attitude is sustained, often without anyone's thinking that it exists. This was brought out at the inquiry in a cross-examination of L. P. Shortis, assistant director of engineering for the reprocessing division of BNFL. Replying to a question by Brian Wynne on plant design as it related to the welfare of employees and social interactions at work, he talked of the need for 'a happy team', and went on to say: 'The sort of thing I had in mind ... [would be] in the detailed consideration of the control of the plant and the control-room layouts and the sort of amenities we have and the sort of training programmes we put our people through and the degree of involvement we want them to have with the venture we are mounting. These are all the human sides of what I have in mind.' The passage is sprinkled with a telling selection of pronouns. Again, the director of health and safety for BNFL told the inquiry that those employed at Windscale receive instruction and training 'to the degree necessary in relation to the nature of their duties, in the hazards associated with their work and the precautions to be taken to provide for the safety of themselves and others'.

But employees may not be consulted in deciding 'the degree necessary'. The official view here may well be that such consultation, given the need for a very large education programme in radiology, would be unbearably time-consuming and expensive. But if safety standards and the nature of occupational hazards *have* been erroneously determined, the employees would have every right to demand that they be invited to the party when safeguards are reappraised. They have, for example, every reason to ask at least for a move in this direction following the evidence given by Dr Alice Stewart on behalf of the Town and Country Planning Association.

In it, she drew on a study of workers at the Hanford nuclear plant in the USA which implied that the risks of contracting cancer in the industry might have been underestimated by as much as a factor of twenty. Her work on occupational medical statistics and in the field of low-level radiation and cancer is respected worldwide, but there is serious doubt as to the validity of the Hanford survey figures, and the study has attracted professional criticism from several sources. Indeed, discussion of Stewart's findings was a last-minute addition to the schedule at the American Association for the Advancement of Science conference in February 1978, where some of her conclusions appear to have been reinforced and others, again, refuted.

But while the controversy continues – while eminent radiologists, epidemiologists and statisticians can dispute the implications of a major research study – it seems reasonable to ask that, where and while there are radiological risks, the employees concerned should be involved in all the stages of assessment and control. There should not be a repeat, for instance, of the failure to record leukaemias and bone-marrow cancers among *former* nuclear process workers, as there has been at Windscale. That omission is being rectified now, and the National Radiological Protection Board has been preparing a National Registry of Radiation Workers which will, eventually, show up any anomalously high death rates among employees; but unions' attitudes may be hardening to the point where *they* provoke such an initiative, rather than wait – somewhat haplessly – for it to be taken by others.

Of accidents in the nuclear industry, it is really necessary to say only that they can and obviously will happen and that their incidence will be that much greater if the scale of operations is enlarged. On the whole, the industry has a rather good record and pays a great deal of attention to engineering design specifically for accident-prevention and accident-control. The accidents that should concern us most, I think, are social and political in character – and they come within the scope of the next chapter.

5 Plutonium and Politics

The most important political issue raised in the controversy over nuclear power is the proliferation of weapons capability that might result from an expansion of civil nuclear activities. Reprocessing of spent fuel would be of particular relevance, since it would make available quantities of plutonium and uranium of bomb-making grade. A fuel-reprocessing plant of the size to be built by BNFL could, each year, recover sufficient plutonium to arm more than 1,000 nuclear weapons if it was reprocessing fuel for thermal reactors, and for 10,000 if it was recycling fuel for fast-breeder reactors. No one denies this, and, since 1977, no one denies either that reactor-manufactured material – persistently claimed by official sources to be unsuitable for bomb manufacture – can be and has been so used.

Controversy hinges on two possible threats. One is the potential of such material for attracting terrorists' attention. The other is that the provision of more civil nuclear facilities and services – notably where agreements and contracts are made to separate and trans-ship separated reactor wastes – will greatly engender the creation of additional nuclear-weapon states. Though this is not the place to argue the need for wholesale disarmament, it ought to be noted that official attitudes in both the Government and the nuclear industry divide – often explicitly – existing possessors of the bomb from non-weapon nations, suggesting that the former may be regarded as more responsible in their likely behaviour than the latter.

Hints of what might happen 'if a country like Uganda got the bomb' have become accepted, therefore – welcomed even – as legitimate cause for concern, and officials in the West have been happy to agree that there is a need for sanctions, constraints, controls and technical barriers to such nations gaining military nuclear strength. This helps to sustain two related doctrines – that the nuclear Non-Proliferation Treaty (NPT) is or can be a satisfactory seal of good peacekeeping; and that the existing nuclear-weapon powers can be trusted to exercise a self-discipline not likely to be encountered in others. The argument fails on at least three counts, one of which is advanced by the pro-nuclear lobby itself.

This, roughly, is that 'if we don't make reprocessing available to countries who want it, then others will' – a position sadly similar to that on the supply of armaments. Reinforcing this is the argument that only by maintaining world leadership in the technologies of reprocessing (and thus the recovery and stock-piling of plutonium) can a country like the UK have an effective voice in framing and fostering sanctions against its proliferation. Dr Tom Cochran, a staff scientist at the Natural Resources Defense Council in Washington, drew a painful historical analogue for the British when he reviewed this argument at the Windscale inquiry. It was, he said, 'strikingly similar' to that used in defence of the slave trade in the eighteenth century. 'In debates in the British Parliament, it was argued that the British could not abolish the slave trade; they could only relinquish it to the Spanish and the Dutch. The British, it was said, should stay in the trade to insure the human treatment of the slaves.' Change *was* possible, though: only a year or so earlier – in 1975/6 – American policy had been to endorse reprocessing as a sensible option. That had now been reversed.

Then there is the question of the NPT itself, which guarantees nothing. It is, essentially, an undertaking by the signatories to be on their best behaviour, consonant with their commercial involvement with civil nuclear projects. It is easily broken or bent to fit national expediency and indeed – in what is perhaps its most important letter – has been a signal failure: there has been virtually no progress towards the general or the nuclear disarmament that is called for in Article VI of the Treaty. The NPT, in any event, does not proscribe or prevent the manu-facture of nuclear weapons, since it does not even define them. A country provided with the necessary intellectual and mechanical skills and given access to reprocessing facilities can, as it were, get to within hours of nuclear-preparedness and not violate the NPT. Most of the non-nuclear components – missiles, delivery systems and associated engineering – can be bought more or less off the shelf. The 'sensitive' parts could be assembled quickly enough at the last moment.

These are simple statements, accepted by a large and distinguished array of international scientists, political leaders and research institutions. They point to an inescapable conclusion: that the deficiencies and loopholes in the NPT and the Safeguards System of the International Atomic Energy Agency (part of the UN) are so great as to render such devices practically worthless. According to the Australian Government's Fox report, the defects '. . . are so serious that existing safeguards may provide only an illusion of protection'. The IAEA, for instance, has no authority, still less the informational resources, to ensure that a nation bent on assembling nuclear military capability does not steadily and clandestinely divert fissile material from its civil power programme to a bomb programme. The possession of plutonium or appropriate isotopes of uranium, especially where the country concerned could demonstrate its need for a breeder-reactor and thus the need for these materials as fuels, only exacerbates the risk.

The British nuclear industry steadfastly refutes these arguments, its spokesmen believing neither in self-denying ordinances nor in the moral imperatives behind unilateralism. BNFL says that it has no evidence that any of the material it has exported under contract (irradiated fuel or its separated products) 'has been used for anything other than its declared end use'. The Canadians could have said much the same up to the very moment India exploded her first thermonuclear device.

There are two additional reasons put forward as to why a British reprocessing plant of the kind proposed by BNFL could not be held to add to proliferation problems. In the first case, it is said that countries denied reprocessing services (i.e. the trans-shipment of their spent fuels and separated fission products to and from European plants) will embark upon the construction of their own facilities rather than risk their nuclear-power ambitions. But developed industrialized nations, like

Japan, do not *want* such facilities, or are severely politically constrained internally and externally from having them. Less developed countries are highly unlikely to be *able* to construct them unless and until the British or the French give them positive technological help as well as the all-important proof that such plants make economic sense.

A second claim is that other countries will perceive the need for and will have the right to pursue a programme of fast-breeder reactors, in which supplies of recovered plutonium, abstracted from *their* thermal-reactor fuels, should be made available to them. But such a programme, again, is likely to be contemplated far less carefully under the present arrangements, where subterfuge, resource-diplomacy, prestige and economic orthodoxy prevail in energy-policy planning, than in a new era of real co-operation and international recognition of the risks of proliferation. If Britain or a major British industrial undertaking is preparing to put its energy eggs in a fast-breeder basket, other countries might scarcely be blamed for following suit.

Professor Joseph Rotblat, Emeritus Professor of Physics at the University of London and founder of the Pugwash standing conference on science and world affairs, was one of the leading witnesses to the Windscale inquiry, arguing – on behalf of the TCPA – that a THORP would be unnecessary without fast breeders and that it would significantly contribute towards the proliferation of weapons. He has written at great length on the subject, but in 1978 refined his work into a five-point recommendation which he urged the UK Government to adopt and make efforts to influence other governments to agree to. The points are these:

(1) Because of its intrinsic links with nuclear weapons, nuclear energy should be taken out of the domain of commercial interests and be tackled as an international enterprise on a global scale.

(2) An International Bank should be established for the acquisition, storage and allocation of nuclear fuels. All enrichment plants should operate under licence from the Bank.

(3) Since recycling of plutonium and uranium from thermal

reactors does not make a vital contribution to the energy resources, but makes the acquisition of nuclear weapons easier, reprocessing should be abandoned and spent fuel stored until further notice. If reprocessing from some reactors is essential, the operation should be carried out under licence from the Bank.

(4) Research and development work on a practical fast-breeder reactor should be carried out as a single project under international auspices and controls. This project is intended to ensure a source of energy if the search for alternative practical sources proves negative.

(5) A massive programme of research on alternative sources of energy, particularly those suitable for developing countries, should be started as a matter of urgency, financed and co-ordinated by a UN agency.

Professor Rotblat's suggestions are excluded from the Parker report and they have, as far as public pronouncements are concerned, received no attention from the British Government. Here, the policy was summed up with hand-wringing honesty by the Foreign Secretary, Dr David Owen, in a speech he made in May 1977. The country could not, he declared, claim to have cut-and-dried answers to the problem of proliferation, which he thought was one of the gravest problems facing the international community. 'The challenge is however such that we are all duty bound to be constantly receptive to new thinking from whatever quarter. We should not be ashamed to admit that, with greater knowledge and awareness, we must progressively tighten up our procedures and sometimes change past policies ... we have made mistakes. International action has been slow, ineffective and insufficient. There have been men of vision, there have been important achievements, but, judged overall, politicians have allowed the urgency and dangers to be swamped by commercial interest and bureaucratic indifference.'

Although there are hints of a need to review the extent to which nuclear power should or may play a part in the total energy mix, there is in the Owen speech little sign of the ferment

– the upheaval in thinking about fundamental nuclear policies – that was by then already going on in the USA and in Britain too. No word about non-nuclear options, and only a sentence or two about the storage of unreprocessed spent fuel. Again, the emphasis was on strengthening NPT and IAEA controls and supporting the International Nuclear Fuel Cycle Evaluation programme, then being set up. Rotblat wrote to *The Times*, after publication of the Parker report, to express his astonishment at the wholesale rejection of the objectors' points, observing that the Carter policy – against reprocessing – had been 'welcomed by Mr Callaghan and endorsed by Dr Owen'. So, in a limited sense, it had; but so had BNFL's application to build a reprocessing plant; and so has the Parker report.

One of the nuclear industry's responses has been, firstly, that civil reactor-grade material could not be used or satisfactorily converted into weapon-grade material; secondly, that weapon-grade material can be adulterated, rendering it similarly unsuitable for bomb manufacture on export to a non-weapon state. Both assertions are unfounded according to much of the evidence given to the Windscale inquiry.

Professor Albert Wohlstetter, Professor in Political Science at the University of Chicago, a consultant to the US Arms Control and Disarmament Agency, and author of the book *Moving toward Life in a Nuclear-Armed Crowd?*, alerted the inquiry to the fact that America had in fact detonated a device armed with reactor-grade plutonium. The phrase 'reactor-grade' is, in Wohlstetter's view, a badly misleading one anyway, since it refers to plutonium recovered from fuels irradiated over the full design period in thermal reactors. For a variety of operational reasons, reactors are frequently run irregularly and over shorter periods and thus discharge spent fuel that is rich in plutonium$_{239}$ – *the* bomb material.

Having, reluctantly, accepted that this may be so, the industry has changed tack to suggest that the yield from bombs made up in this way would be 'only' a few kilotons (i.e., the equivalent of a few thousand tons of TNT explosive) and that a country so equipped would, in the words of BNFL's deputy managing director, Dr Donald Avery, hardly be a 'genuine nuclear force'.

Wohlstetter says: 'I suppose this is a matter of definition. Dr Avery's definition, however, is unlikely to be accepted by a country without nuclear weapons facing an adversary with such weapons. In 1945, the US had only two nuclear weapons in the low-kiloton range' (developed beyond yield expectations in only seven months), 'and these were enough to bring quite a large war to conclusion.'

The industry's argument – advanced apropos of reprocessing constraints – that a determined country could and would do what it considered necessary, without technical assistance from elsewhere, is therefore a catch-all. As for adulterating, or 'spiking', returned fuel so that it cannot be used for bomb manufacture – broadly by irradiating it specifically for overseas reactors – the commercial and technical complications appear to be extremely unfavourable to the supplying company or country. It would involve the use of additional and costly plant and would, if effective, reduce the energy content of the returned fuel. The so-called Civex process, more recently canvassed by the UKAEA as a sort of super-spiking, in which fuel would be made so radioactive as to frighten off direct theft and diversion between countries, raises just as many technical problems concerning authorized handling, and is, in any event, not held in prospect for at least fifty years.

What, then, of the direct theft, diversion and use of nuclear materials by terrorists, guerrillas and other non-state groups? There are three sources of threat, and they are all taken seriously by government and industry alike. First there is the group who would seek to acquire a tactical weapon or the means of its manufacture. Then there are those who would breach nuclear security to damage plant or equipment or, simply by their action, cause serious political and social dislocation. Lastly are those who might mount a similar operation to *prove* that the first two threats are irresistible: in other words, an extreme and militant section of the antinuclear lobby.

So subtle and pervasive are the requirements of a security system charged with averting such threats that a great many people outside these groups are inevitably to be regarded as risks. They include those who work in and with the nuclear industry, who will be subject to surveillance and vetting. And they include those who attend meetings or public gatherings where the subject is discussed. Though it became something of an open laughing matter at the time, the presence throughout the Windscale inquiry of two officers from the Special Branch indicates official concern about the third category of potential risk. An assurance was given during the inquiry that security investigations would not be made about any person unless 'subversion' was suspected. The word was not defined, does not exist in English law and can – one may reasonably suspect from examples in the recent past – be an all-purpose description for almost any expression of opposition to official policy.

The threats to nuclear security have been characteristically well documented in the USA, where the Nuclear Regulatory Commission prepared in 1975 a feasibility study on the need for a new Security Agency. This followed an earlier study which described 'problems and shortcomings' of existing safeguards. The NRC report said:

'Threats to nuclear facilities and material can come from external or internal sources. External threats would include overt acts of theft or sabotage. They span a scale ranging from mischief and minor nuisance through co-ordinated attacks, which at some point would take on the character of civil war. Internal threats are most often postulated as being covert and might involve diversion of material, the perpetration of hoaxes and, perhaps, sabotage. They span a scale from minor pilferage by individuals, through collusion, all the way through revolutionary conspiracies, in which entire plants might be covertly controlled.'

The NRC prepared a simulation exercise to estimate the 'maximum credible threat', and concluded that, in the case of external risk, the action would probably be taken by a group of between six and twelve persons, and that, for internal threat, they would be dealing with two or three individuals in collusion. The Office of Technology Assessment, using data provided by the Rand Corporation reviewing commando tactics in military history and even analysing the English Great Train Robbery, independently concluded that seven to fifteen attackers would present a credible threat. Even more agencies became involved as these studies proceeded – the FBI, CIA, Department of State and Defense Intelligence had all been consulted by 1977, when the appropriate NRC Task Force concluded that '. . . there can be no assurance of detection of this level of threat [by a group of ten to twelve dedicated, well-trained and well-equipped fanatics] prior to an attempted malevolent act'. No assurance of detection could be given 'unless group sizes become very large, that is "army size"'.

Several studies have indicated strongly that terrorists who did steal fission materials could, without supreme difficulty, construct a 'credible' nuclear explosive. Again, the OTA Task Force consulted five weapons experts for advice and found that '. . . a small group of people, none of whom have ever had access to the classified literature, could possibly design and build a crude nuclear explosive device. They would not necessarily require a great deal of technological equipment or have to undertake any experiments. Only modest machine-shop facilities, that could be contracted for without arousing suspicion, would be required. The financial resources for acquisition on open markets need not exceed a fraction of a million dollars. The group would have to include a person capable of searching and understanding the technical literature in several fields, and a "jack of all trades" technician.'

The 'fraction of a million dollars', according to Tom Cochran of the NRDC, could be as little as $1,000; the researcher could also be the technician; the device could be armed with about 3 kilograms of plutonium (which would arise in tonne-quantities in a thermal-oxide reprocessing plant). There *would* be a chance that the device would fail or that a mishap in its construction would injure or kill one of the group. That is unlikely to deter fanatics of the PLO, Baader Meinhof, IRA or Red Army type. But what is also clear from the now large library on this subject – and from sensible inference – is that fissionable material, once stolen, need not be used at all. Indeed, if all a terrorist group did was to leave signs that it had entered a plant, the disruption that would follow leaves little to be imagined. There would be no waiting around while inventories were checked to see if plutonium or other material really had been stolen (not an easy thing to do in a hurry, and still less so if material is thought to have been diverted over a period of time). Security crackdown would be immediate and probably draconian.

Plutonium is thus, in the words of Louis Blom-Cooper, who appeared at the Windscale inquiry for the National Council for Civil Liberties, 'a symbolic, newsworthy, and overwhelmingly fear-inducing target for terrorist action'.

Recognizing this, what responses are likely from official quarters? The evidence – well documented in Britain by Michael Flood and Robin Grove-White in their study *Nuclear Prospects – a Comment on the Individual, the State, and Nuclear Power* – is that they have a potential social impact no less than that of the Second World War. The imposition of controls, the withdrawal of rights, and the sheer deployment of special powers needed effectively to thwart nuclear terrorism are, quite plausibly, equal to those that operate in a country at battle. Already, there is a heavy security system, operated under various Acts, covering the civil nuclear power and reprocessing programme. All professional staff (and some so-called industrial staff) at UKAEA/BNFL establishments are 'positively vetted' before appointment. This involves a close scrutiny of their personal lives to reveal political sympathies and associations and also to monitor possible character defects which might make them vulnerable to blackmail, political seduction or even plain nervous breakdown. Periodic and detailed checks are made on employees, invariably without their full knowledge. Trade-unionists within the industry are watched with particular care: Flood and Grove-White note that it is one of MI5's functions to monitor 'the instigation of strikes . . . in order to serve the purpose of some foreign power'.

No more than seven to eight tons of plutonium is currently handled each year in Britain between Windscale and the proto-type fast-breeders at Dounreay, but the risk of any being abstracted for use either as an explosive device or as a lethally toxic aerosol has led to the arming of a private police force, under the Atomic Energy (Special Constables) Act of 1976. This force is answerable not to the Home Office, but to the UKAEA, and has the power of armed pursuit. In the words of Justice – the British section of the International Commission of Jurists – it is constitutionally unique: '. . . its structure conflicts with all our traditions or civilian and politically accountable policing'.

But what is forecast by the nuclear industry itself must be seen as a veritable invitation to a police state. With 100 or so reactors in the early decades of the next century, half of them FBRs, and 400 tons or more of plutonium fuel each year to be shipped between reactors and reprocessing plant, a colossal surveillance and security network would be required, and it would have to cover not merely employees of the UKAEA, BNFL and the Generating Boards, but also those in industries and companies providing goods and services. The list is virtually limitless, for it would have to contain families, friends and acquaintances if it were to be plausible. The vetting would embrace contractors on building sites, telephone-exchange operators and local shop-keepers, and would prove especially rigorous for those involved with transport – seamen, railway workers, dockers and their respective unions and union officers.

But it is on the third category – the antinuclear lobby and its supporters – that the effects, if taken to a logical conclusion, will be most insidious. Already the movement has been openly associated with political extremism (by an author who shall remain unnamed but who had to retract his accusations, by the EETPU – the electricians' and plumbers' union – and by a witness at the Windscale inquiry). It is not hard to envisage that the lobby would attract security attention disproportionate to its actual numbers. Louis Blom-Cooper warns that members of organizations concerned to protect the environment would soon come under secret surveillance: 'Files will be opened on opponents of nuclear development: they will be subjected to telephone taps, mail interference, perhaps opening of their bank accounts, and other invasions of their privacy. Political demonstrations will be monitored.' From here it would be a short step to stern censorship or opposition voices, with curtailment or outright withdrawal of the right of objectors to meet, discuss, air or publish.

To go back to the difficult question of 'subversion', how *is* it to be interpreted? Lord Denning, in his Profumo report of 1963, took the word to mean contemplation of 'the overthrow of the Government by unlawful means'. Major-General Kitson, in *Low Intensity Operations*, saw subversion as involving '... illegal measures to overthrow those governing the country at the time', and added: 'or to force them to do things which they do not want to'. Even that somewhat bizarre enlargement of the definition, which also includes 'the use of political and economic pressure, strikes, protest marches, and propaganda', is, as Blom-Cooper observes, restrained when compared with the one given by Brian Crozier, Director of the Institute for the Study of Conflict. For him, 'subversive intent' would be signalled by, among other things, 'the discrediting of capitalism' and 'the denigration of national achievements'. Blom-Cooper concludes that 'one man's subversion is another's bedtime reading'.

The issue remains a deeply serious one. If the antinuclear movement grows, and there seems every reason to believe that it will grow, the state's security response will intensify along with it until one of two things happens: official nuclear policy is changed; or civil disobedience and insurrection prevails. As we see in the next chapter, normally quiet and passive people have threatened militant action in one part of Scotland. If a large reactor-building programme were to be put into operation, such militancy could be expected to erupt in many places. Normal quieting tactics – the public inquiry, for one – would be unequal to the sheer numbers involved.

What the Windscale tribunal, BNFL and spokesmen for Government agencies involved do not appear fully to appreciate is the extent to which these fears are profoundly related to other parts of the antinuclear case. Parker, in his report, sees a problem in enhanced surveillance, but appears to make a straight choice between the need for security on the one hand and a reduced standard of living (i.e. with reduced nuclear power or none at all) on the other. BNFL sees that security measures 'may be distasteful to some' but finds them related to the state of society at large and not to the growth of nuclear power. Parliament did not relish the creation of the nuclear Special Constables, but created them all the same. What they all miss is an ecology of argument among the more thoughtful objectors which sees the extension of nuclear *and* the extension of security powers as two undesirable effects of the same cause: technocentricity.

The only open and 'independent' survey of public attitudes to nuclear power in Britain was that conducted by the magazine *New Society* in March 1977. I use quotation marks without intending offence: no survey based on a structured questionnaire is ever independent. What emerged from the survey was a picture of confusion and contradiction, with a quarter of the respondents claiming that they would do everything in their power to resist the siting of a nuclear plant near their homes, some of them voting with the 49 per cent who declared themselves to be broadly in favour of an expansion of nuclear capacity. The fact is that no one can yet gauge the potential opposition with any accuracy. As Peter Taylor, of the Oxford Political Ecology Research Group notes, an adequate study would require immense preparation, and it would need to determine how well informed people are about the benefits and liabilities of nuclear power.

It may be that the acceptability of nuclear power is even less measurable by ballot than are other issues. Looking at the *New Society* study, Taylor points out that a very large group within the general population is strongly opposed to nuclear power for reasons that apparently include intuitive or emotive reactions: 'I believe it is important that we try to identify the nature of these reactions. If we do not know what "the environment" means to people, we cannot assess an "environmental impact"; if we do not know how people view the risks of technology, then we cannot define what is an acceptable risk. If we try and we get it wrong, then we may promote conflict – besides, there is the political and ethical question of participation in all these deliberations.'

Which is just another way of saying that public opposition, rightly or mistakenly based, should be one of the factors considered in taking decisions on major issues such as the nuclear one. 'We shall not be moved' is a largely emotive response, but it is nonetheless a politically real one, and to this is allied a new-found facility for challenging the expert view, the Government decision, the prepackaged plan. Nowhere in Britain has this been vindicated more forcibly than in the movement to question and resist the construction of motorways, trunk roads and major urban highways. After only seven or eight years of campaigning, the objectors were rewarded in 1978 with the Leitch report, which roundly condemned governmental tactics for formulating and prosecuting official road-building schemes. The opposition to nuclear power is likely to concert itself in much less time.

For a start, the controversy has immediacy and specificity. Opposing the building of a series of roads, even if questions of general, national need were admitted (which they were not), is a much more difficult task than objecting to a single, rather small chemical plant, site preparation for which is in train as the objections begin. Other attitudes have changed in the meantime: a petition of 27,000 signatures against the BNFL plan was raised in a matter of days in *one county*. They were, figuratively and effectively, votes on a national issue and not a local one.

Then, with the procedural freedoms conferred upon the Windscale inquiry, it became an international issue, open to challenge on the widest possible brief. In conceding these freedoms – some would call them rights – the British Government has uncaged an animal whose behaviour is highly unpredictable in all but one respect: it will not allow itself to be put back in the cage. That in itself does not tell us what we need to know about public acceptability. Are the objectors a vociferous but unrepresentative nuisance? Are they a groundswell of much wider opinion, as yet unheard? Or are they – and this seems more likely – a relatively fortunate section of the public who have had the education, the time and the resources to develop and articulate a countervailing response?

6 Chain Reactions

The Windscale inquiry may not have given environmentalists the answers they wanted, but it provided them with an exceptional forum in which to ask the questions. For many of those who appeared as objectors, it was the first opportunity to say in public what they had been saying among themselves for years. It would be crude and misleading to ascribe to them a common view, but a broad consensus undoubtedly existed – that the social and environmental penalties to be paid in a society based on the use of advanced technologies are high and becoming intolerably so. The prospect of a growing commitment to nuclear power around the world has been both a foodstuff and a focus for these fears, which I shall try to elaborate.

What of the nuclear ' debate' itself, though? Is it a debate at all – or has controversy passed the point where the various issues might be handled according to reason and logic? At the beginning of the seventies, there was no concerted challenge to the wisdom of going ahead with programmes of electricity generation from nuclear power. There was *concern* – notably in the USA – about the risks involved in building and operating fast-breeder reactors, and a handful of people there were ready to oppose the construction of so-called nuclear parks. American energy planners were talking then (and some still are) of having 4,000 small nuclear stations sited around the country's seaboard, and local protest was widespread.

Since then, attitudes have changed dramatically. No longer is opposition of a not-on-*our*-doorstep-please variety the norm: it has given way to a generic international campaign – in keeping with other reformist missions such as those concerned with resource conservation, wildlife protection and pollution abatement. And where doubts about nuclear power were expressed only by a handful of heterodox economists, dons and disaffected engineers, they are now embraced by people from virtually every social group – bishops and beggars, premiers and pacifists, students and statesmen, Tories and Trotskyists. They do not include, in Britain, official members of the Communist Party (though CPs in other parts of Europe take an antinuclear stance); just as surprising, the objectors have a small number of sympathizers within the industry itself.

What the Windscale inquiry did, more than anything else perhaps, was to unify the opposition, to create a more or less well-defined camp to which one either belongs or does not. During the inquiry and since, the sense of crusade has been reinforced by the number and variety of antinuclear moves elsewhere in the world. They range from acts of rebelliousness to national policy changes – and they almost all suggest that the days of genuine debate, if they ever did exist, are over. The protagonists are now engaged in what one might call emergency politics. Some have likened the mood to that of twenty years ago, when the cause of unilateral disarmament brought scores of thousands out on to the streets of the Western world in opposition to their governments' nuclear military policies.

The comparison cannot be sustained. The CNDs were soft and relatively insular: their largely youthful supporters saw civil disobedience and mass rallies as means to an end that had little to do with the other social, economic and technological issues that touched their lives. The idealism gave out and their energies were absorbed into living and working in a society entering a period of unparalleled material prosperity. Later came renewed disaffection, and it was increasingly all-embracing: anti-militarism, concern for racial and sexual equality, compassion for an underfed and exploited Third World, demands for open and accountable government, and fears that industry was brutalizing man and poisoning his natural environment – all these came together in one large but unnamed movement.

Unnamed, that is, if one is thinking about the informal gathering of these issues. Officially, the United Nations organization has an interest in them too, but the UN specialized agencies have been notoriously ineffectual in all but a handful of cases.

The initiative now is with a loose federation of groups and individuals for whom the growth of nuclear power is emblematic of much that they abhor. For them, Gandhian tactics of opposition have not been and will not be characteristic. Instead, direct physical and political action is going to be the norm. The Windscale inquiry itself, though conducted with a certain stateliness, was a creature of intense lobbying, and throughout the hearings there was a strong feeling of its being a prelude to the more radical expression of opposition.

Four of the objecting groups represented at the inquiry indicated that they favoured or foresaw direct-action campaigns in Britain. Those campaigns are directed towards three distinct fronts. The first, a broad one, is a challenge to nuclear power in general and to the reprocessing of spent reactor fuels in particular. Along this front would be marshalled most, though not all, of the objectors to THORP. Friends of the Earth, for instance, would at the present time challenge outright only the reprocessing issue, whereas the Society for Environmental Improvement and the Yorkshire branch of the National Union of Mineworkers both believe in and predict militant action in pursuit of phasing out the whole nuclear-power industry.

In an attitude survey carried out for the SEI in sixteen locations throughout Britain, 500 out of the 1,200 polled indicated that they would be prepared to demonstrate against the construction of a commercial fast-breeder reactor, which – at the time of writing – is to be the subject of another major inquiry, the terms and timing of which have yet to be announced. This, the second of the fronts, has almost universal support among antinuclear groups both in Britain and abroad, where it is widely felt that FBRs are an unnecessary and highly dangerous addition to the power-generation inventory. Indeed, opposition to the FBR was at the back – and often the front – of many submissions made to the Windscale inquiry.

The third front, and one on which real and early militancy can be expected, is that concerning the burial of highly radioactive reprocessed wastes. In south-west Scotland, across the Solway Firth from Windscale, the United Kingdom Atomic Energy Authority has applied for planning permission to drill for core samples of the rock 500 metres below the Galloway hills. Here, if the deep granite strata were found to be structurally sound, seismically stable and hydrologically impervious, could be one of the sites where the most toxic of the wastes from European and Japanese reactors might be stored. 'Disposed of' is the phrase preferred by the industry, which is investigating other sites in Britain, France, Germany, Holland, Belgium and Italy.

Opposition in Scotland has been characterized as reflex nationalism lighting on an easily promoted popular cause. Protest expressed elsewhere, however, suggests that this is not so and that very real fears – substantiable or not – have been aroused about the threat to environmental health and integrity posed by the HARVEST plan to seal high-level effluent into glass blocks and plant them underground. The process for vitrifying the wastes is, anyway, at an uncertain development stage; safe physical isolation for some of the isotopes in the waste would need to be guaranteed for periods ranging from 600 years for fission products to half a million years for by-products, such as $plutonium_{239}$.

Guarantees of this kind, of course, are impossible to give. And what, for the objectors, compounds the unsatisfactoriness of going ahead with schemes of vitrification and underground storage is the wide diversity of specialist opinion on the likely chemical, mineralogical and geological behaviour to be anticipated at the burial sites. As with so many aspects of the nuclear controversy, you pays your expert and you takes your choice. Many opponents feel no need of scientific consultation on the matter: theirs is a gut belief that this is not 'disposal' at all but a short-term shelving of a problem which lays an unacceptable burden on the natural environment and bequeaths the difficulties of nuclear-waste management to generations who, for all that can be predicted, may have lost the science of radiological protection but who might badly need it.

The fact is that no one wants the wastes on (or beneath) their doorstep. On more than one occasion, spokesmen for the industry have gone so far as to concede that if the most technically suitable sites were near or at major centres of population, they would not be contemplated for fear of massive challenge.

But the fierceness of opposition in rural Scotland has more than matched the smallness in numbers. Opinion polls there have produced unprecedented response and unanimity against the storage proposals, and in the spring of 1978 people in Galloway – a passive, pastoral, peaceable community – were talking of guerrilla action, of their readiness to go to prison to prevent nuclear dumping. A local medical practitioner went further: 'Democracy is at stake, and people are prepared to die to stop this.'

Similarly concerted opposition has been voiced towards the mining of uranium in Britain. In the Orkney islands – politically part of Scotland but traditionally very much an independent settlement – public feeling ran so strongly that the South of Scotland Electricity Board abandoned a programme of prospecting for uranium, partly financed by the European Commission. On Deeside, where uranium traces have been detected, the Board met with equally strong opposition: the SSEB's chairman confessed that he had 'underestimated the reaction the proposals would arouse'. It is perhaps significant that both in the Orkneys and on Deeside there has been comparatively little hostility on environmental/aesthetic grounds to the exploration for and exploitation of North Sea oil, with which both localities are very closely involved.

On the other side of the world, in Australia, uranium mining has become a major political issue. Trade unions there, already well known for imposing sanctions in the construction industry against projects deemed to be environmentally or socially unacceptable to members of the labour movement, mobilized thousands of workers in support of a 'black ban' – a boycott – on the export of uranium ore from Queensland. The deputy leader of the Opposition warned trade-unionists that there was the danger of 'extreme penal provisions' being invoked against them, hinted that the Government was preparing to insinuate *agents provocateurs* in the protesters' ranks and likened the conflict to that which took place during Australia's involvement in the Vietnam war.

English dockworkers, responding to Australian union requests and to local environmentalists' suggestions, prepared to block imports of any ore that might be clandestinely shipped from Queensland. The industry reacted by bringing in ore from Niger aboard D C 8 freighters – the first time, it is believed, that ore has been carried by air. Action of this kind, clearly, is not one of the components of debate – any more than are the threats by British coalminers to picket the nuclear industry's factories and facilities and block the transport of fuels and wastes. Militancy and secrecy feed insidiously off themselves, with entrenchment, provocation and intensified conflict the principal results.

In America, a further challenge to the nuclear industry has come from the successful contesting, in a federal district court, of the constitutionality of the Price Anderson Act, which limits the liability of a public utility in the event of a disaster. If this judgement is sustained, on appeal, the size of potential damage claims would make it all but impossible for utilities to obtain insurance. Amory Lovins, for instance, has estimated that an accident involving the failure of a reactor containment structure could lead to 10,000 deaths and property damage of the order of £10 billion. The antinuclear movement in Britain has noticed the wholesale exclusion of risks arising from nuclear accidents in common insurance cover. A typical mortgage protection policy (my own) contains the following clauses – inserted on the advice of the British Insurance Association, whose actuarial intelligence naturally has some standing:

This policy does not cover loss or destruction of or damage to any property whatsoever resulting or arising therefrom or any consequential loss; any legal liability of whatsoever nature; or any bodily injury – directly or indirectly caused by or contributed to by or arising from –
1. ionising radiations or contamination by radioactivity from any nuclear fuel or from any nuclear waste from the combustion of nuclear fuel
2. the radioactive, toxic, or explosive, or other hazardous properties of any explosive nuclear assembly or nuclear component thereof.

Looking for protection nevertheless, an FoE member asked his insurance company whether it would separately indemnify him and his family against nuclear contamination. He was told they would not. Pursuing his inquiry, he asked why not. The risk, he was told, was *too small* for a policy to be worth his while. Not surprisingly, there is widespread concern about what is clearly a breakdown of normal compensatory law.

In France, where the nuclear conflict has been at its most bitter and violent, there are few signs that the Government wants to exercise its powers of mediation and restraint. Not for the French a programme of public inquiries: so far, official reaction to protest has been one of singularly brutal and heavyhanded repression. In spite of the serious technical and economic uncertainties which hang over it, the French energy planners are determined to pursue the goal of producing 70 per cent of the nation's electricity in nuclear stations by 1985, after which the country's thermal reactors would gradually give way to the world's most ambitious network of FBRs. The first of these, at Creys Malville, will mark the spot where the nuclear controversy claimed its first fatal victim.

Opposition to this FBR came to a head first in 1976, when protesters staged a non-violent occupation of the site. The demonstration was broken up by riot police: tear-gas and truncheons were used, and several protesters were injured. Almost exactly a year later, in July 1977, the site was besieged by demonstrators from all over Europe. Several left-wing groups had made plain their intention to take part; others, it is thought, had been clandestinely monitored – particularly in Germany. The French authorities, seeing a potential repeat of the anti-Government student outbursts of 1968, reacted with what amounted to a full-scale counter-insurgency plan. An area of ten square miles was sealed off, roads around Lyon were blocked, and thousands of military and civilian police were marshalled and equipped with armoured vehicles and automatic weapons.

Objectors' groups and trade unions – which a few months earlier had reappraised their position on nuclear power when the ecologists took 10 per cent of the French municipal-election votes – had repeatedly undertaken to keep the protest at Malville a legal and non-violent one. The local department Prefect, however, warned that anyone who tried to get inside the cordoned area did so at their own risk: the fast-breeder site, he said, was 'a national asset and must not be damaged'. It was not –

but the country's reputation probably was. A young unarmed biology teacher, Vital Michalon, was killed by a percussion grenade as police and troops stormed the crowd. Brigadier Fernand Touzeau, who ordered the use of grenades, lost a hand as one exploded prematurely. Many people were injured and taken to hospitals.

The year saw similarly vehement demonstrations in Germany and Spain. Continuing opposition to plant developments at Brokdorf and Aschendorf brought clashes between environmentalists and police. In Europe's newest democracy, official reaction was quieter, though conflict in the widest sense was already open. Shortly after the Malville incidents, 200,000 demonstrators took part in a march in Bilbao, appealing to the Spanish Government to halt construction of a nuclear plant at Lemoniz. Organized by the Commission for the Defence of the Basque Coast against Nuclear Pollution, they too included many socialists and communists in their ranks as well as the separatists: the antinuclear cause was seen as a unifying political movement for many of the groups emerging after the end of the Franco regime. But with Spain planning to be, by 1985, the third largest nuclear-energy producer in Europe – behind France and West Germany – the chances of a peaceable resolution of conflict seemed low.

In Britain and the United States, opposition to the use and expansion of nuclear power is largely from a semi-institutionalized middle-class movement. Apart from some special-interest groups (like coalminers) and people living close to proposed plants or nuclear sites where local employment is in balance, feeling has not been politicized to anything like the same extent as in mainland Europe, Japan or Australia. In Sweden, for instance, the Government changed hands in 1976 with nuclear power a significant voting issue. The Dutch Labour Party aligned itself with the antinuclear movement early in the seventies – though it has proved unable, as party to a coalition government, to turn its views into official policy. In Japan, the environmentalists, though generally parochial in their concerns, claim the support of Sohyo – the country's biggest of the labour federations – and Gensuikin (the Japan Congress against

Atomic and Hydrogen Bombs). Together with other citizens' associations, some 15 million Japanese might be numbered among those opposed to nuclear development, and as Japan's electric-power utilities are important potential customers for the supply and reprocessing of fission materials, serious conflict seems likely.

Even in less volatile countries, the political problems raised by nuclear power are beginning to disrupt energy programmes and parliamentary business alike. By the beginning of 1978, Austria had joined the list, with the country's first nuclear generating station, at Zwentendorf, held up while public discussions were arranged and promises given for a full-scale Government debate. Other nuclear projects in Austria are, for the time being, scrapped. Though there *is* union support for nuclear energy there, Chancellor Kreisky, who lived in Sweden for twelve years, is said to be 'keenly aware of the explosive political implications of nuclear-power controversies'.

In Sweden itself, the argument has gone full circle. In January 1978, a committee appointed by the Government reported that future energy supplies could not be guaranteed without a nuclear programme – which Prime Minister Thorbjorn Fälldin, before his election, had sworn to phase out by 1985. A month later, with the Cabinet bitterly divided on the question, it was agreed to put off any decision and withhold funds for expanding the country's nuclear-power programme. The controversy, said Mr Fälldin, was a major threat to Government unity.

Another, albeit small, political administration caught in the crossfire is that of the Channel Islands. The parliaments of these semi-independent British states have found themselves petitioned by one in ten of the 120,000 population. Newly formed groups there, partly encouraged – if that is the word – by antinuclear activists in Normandy, have voiced their fears over the French reprocessing plant at Cap la Hague, only fifteen miles across the water, and at plans to build a twenty-station nuclear park near Cherbourg. Accusations that the groups were politically motivated only served to inflame the objectors' feelings.

Not far away, opposition has also been growing in the Republic of Ireland and in the Isle of Man, both of them closer neighbours to Windscale and the stations along the English west coast and Wales than is London. Here again, the politics of independence and separatism are an important part of the controversy, though these states are, for the moment, concentrating on the environmental risks posed by possible radioactive contamination of the Irish Sea.

Nuclear conflict is almost as exclusively the property of Western industrialized countries as are the nuclear initiatives. Britain and the United States, concerned mainly about the proliferation of nuclear weapons and weapon-making capability in *others'* hands, have spent much of their diplomatic energies in trying to make the ten-year-old nuclear Non-Proliferation Treaty work. Of the known nuclear-weapon states, they and the USSR have signed and ratified the NPT; China and France have not. One major uranium supplier, Canada, has ratified the treaty and has tightened the safeguards in its conditions of sale; another, South Africa, is not a party to the NPT. Beyond this list is an increasing number of countries – roughly three more every two years – who are operating or constructing nuclear plant.

Growing concern over the potential in these countries for manufacturing and fabricating military nuclear hardware is matched, in the West, by the commercial and political transgression of some of the terms of the NPT. On the one hand, there is alarm at the prospect of nuclear-arms development in countries such as India, Pakistan, Cuba, Israel, Argentine, Brazil and South Africa. On the other is the relentless pressure to sell nuclear equipment and materials in world markets. Morally, the weapon states are on shifting sand, having failed to uphold one of the NPT's most important articles, which calls upon the parties to initiate general and complete nuclear disarmament. It is an old theme: do-as-we-say, not do-as-we-do. But in 1976 the USA did at least take an overt and strong stand.

In October of that year, President Ford announced that the country should no longer regard the reprocessing of spent fuel – and thus the production of plutonium – as a necessary or inevitable part of the nuclear-fuel cycle: '... avoidance of proliferation must take precedence over economic interests'.

The philosophy has been hardened considerably since President Carter took office, notably with his statement in April 1977, which represented a dramatic policy shift and which was central to much of the debate about Windscale. It was also in stark contrast to well-intentioned but faint-hearted statements on the subject from the British Government. Carter listed seven reforms intended to limit the spread of weapon-making capability and based on a review that he began on his first day in the presidency.

Accordingly, the USA deferred indefinitely the commercial reprocessing and recycling of plutonium from American power reactors. The existing nuclear-power programme could, said Carter, be sustained without such reprocessing, and America's own THORP – at Barnwell in South Carolina – would not be given the funds necessary for its completion. The country's FBR programme was also deferred. Research and development projects were to be aimed at the limitation of access to weapons-grade materials, and there would be a strengthened embargo on the export of equipment or technology for enriching uranium and reprocessing spent fuel. At the Downing Street Summit conference, when Carter paid his first visit to Britain, it was announced that an international nuclear-fuel-cycle evaluation programme (INFCE) would be set up to take the Carter initiative into a practical realm.

INFCE, a two-year technical study, will not commit anyone to anything, but it is going on in parallel with diplomatic moves in some of the more sensitive parts of the world. US Secretary of State Cyrus Vance made a whirlwind three-country Latin American tour in November 1977, and the question of human rights – which most observers had anticipated would be the centrepiece of his discussions – became a side-issue. In Brazil and the Argentine, the Carter policy on nuclear controls dominated the Vance talks. Although he extracted a promise from the Argentinians to ratify the NPT, the Brazilian nuclear programme was actually being unfettered as contracts with American, British, Dutch and German suppliers were negotiated. And American overtures to the Governments of India and Pakistan have met with even more uncertain results.

Pakistan is determined to have a reprocessing plant and in 1976 signed a deal for the necessary equipment to be bought from France. This would, inevitably, be partly financed out of development-aid funds – of which those from the USA alone are now running at about $80 million a year. This would be cut off if Pakistan were to press ahead with its plan. France, unwilling to go back on its commercial word, is nevertheless a member of the fifteen-nation Nuclear Suppliers' Group, which has put the reprocessing plant of the sort Pakistan wants on the so-called 'trigger list' of proliferation-prone items. The solution offered by France is to supply Pakistan with a co-processing plant, in which the oxides of plutonium and uranium are mixed and from which bomb-grade plutonium is very difficult to extract.

Pakistan has refused this solution, saying that the technology of co-processing is unproven. The country's military ruler, General Zia-ul Haq, accused the Americans of putting discriminatory pressure on the country. The Middle East and Pakistan badly needed reprocessing facilities, he told a press conference in January 1978, and though he hastened to assure his audience that the technologies would not be passed on to 'one of the more radical Arab states', there is a real fear in the West that this would be a consequence. His predecessor, Zulfikar ali Bhutto, implicitly acknowledged that Pakistan's purpose in separating plutonium in a reprocessing plant would be to develop military nuclear facilities. The French deal, he said early in 1977, could 'easily be cancelled' – but only if the nuclear powers would destroy all nuclear weapons. A persistent rumour in Pakistan had been that he was able to obtain funds from Arab countries to finance nuclear projects and even that he had undertaken to supply the bomb to Arab nations to match the Israelis' supposed nuclear capacity.

Perhaps. And perhaps not. But the perhapses are what President Carter's non-proliferation initiatives are designed to remove. Unfortunately, they come at a time when many countries in the Third World are contemplating substantial industrial development programmes in which – *by example from the West* – nuclear power is expected to play a significant role. And a country determined to have the technology, whether for civil or military use, or both, has so far been able to do so: India is a prime example.

The explosion by India, in 1974, of a thermonuclear device was probably the most important single factor contributing to the redoubled concern over proliferation. The United States and Canada, from which India obtained the essential materials, were gravely embarrassed, and after a long series of comparatively fruitless representations to the Indians decided on a hard line. At the beginning of 1978, President Carter, breaking with most traditions of statesmanship, told the Indian Prime Minister, Morai Desai, that he would have to sign the NPT 'or else'. Carter talked of sending a 'cold, blunt letter' from Washington if agreement was not secured. A couple of weeks later, British Prime Minister Jim Callaghan appeared to have smoothed over the gaffe and have procured India's agreement to relax its attitude to international inspection of the country's nuclear installations.

Diplomatic pressure of this kind may be counterproductive, even where it is reinforced by another suite of arguments against the development of nuclear-power programmes in the Third World. These are embraced notably by the International Bank for Reconstruction and Development (the World Bank) which will now no longer make funds available for nuclear projects in the poorest countries. Without national electricity-transmission systems or the possession of modern electrical appliances to plug into them, the Bank argues, it makes no economic sense to go for ambitious generating stations – especially fission-power plants; the emphasis should be swung towards technologies more suited to the needs and skills of impoverished and predominantly rural nations like India. Agencies like the IBRD and indeed some in India itself are interested in backing appropriate technology (AT) projects instead, making the maximum use of local materials, fuels, labour and expertise.

They are opposed by a desire for technological prestige, which runs high throughout the underdeveloped world. India, though poor, is officially a very industry-conscious country. It educates more scientists and technicians, for instance, than any other countries with the exception of the USA and the USSR. And it is just as keen to emulate the West in its strategic capabilities. So arguments about non-proliferation fall, if not on deaf ears, then certainly on unsympathetic ones. The issue, one should add, is paralleled in most other aspects of economic and social development – notably in the Third World approaches to transport, urban planning and agricultural production. The West's is the model, and having run hard to catch up, philosophically, with that model, it is not surprising that they are unwilling to let it slip from their grasp.

But India does at least have the freedom to discuss the issue. In countries where democracy has been killed or curtailed, the nuclear question is not debated. Iran, the USSR, parts of Latin America and South Africa are the leading examples, but we should also include Israel and the Philippines, Mexico and South Korea, all of whom are potential nuclear military states.

Thus these countries and others are involved in a related but separate part of the ferment now opening up in Europe and the USA. The paradoxical fact is that Western governments have stolen some of their environmentalists' clothes in arguing that the Third World is an inappropriate market for nuclear power. Sir John Hill, chairman of the United Kingdom Atomic Energy Authority, freely conceded in an energy forum at the Royal Institution in October 1977 that the case for nuclear-electric projects in the underdeveloped world was refutable. 'None of us thinks it makes any economic sense at all; this is not the way the Third World ought to develop at this stage ... the developing countries must concentrate on coal, oil and biomass ...' Similar sentiments had been expressed about the *developed* world by objectors at the Windscale inquiry.

To the astonishment of some British MPs and the nuclear industry, their Government decided in the spring of 1978 to make public a letter written to the Foreign and Commonwealth Office by Dr Joseph Nye, an Under-Secretary at the US State Department. The letter, sent in December of the previous year, came – in the words of one commentator – 'as near as diplomatically possible' to telling Britain not to proceed with the Windscale expansion proposal. It pointed out that President

Carter was opposed to THORPs on two grounds: their possible contribution to the greater risk of proliferation; and a general belief in the USA that such plants were not needed. This exchange, emerging at about the same time as inspired and speculative predictions that Windscale was to get the go-ahead, underlined the fact that the West itself is riven on the subject.

Even in France, the least conciliatory of the industrialized nuclear nations, the issue has lately become a worrying one for the major political parties. Both the Gaullists and the Communists have been – and mostly still are – staunch supporters of nuclear power. But they nevertheless jointly endorsed a parliamentary committee's report which voiced grave concern on the economic case for a large nuclear electricity-generating programme. The report noted that the cost per unit of nuclear energy had trebled in four years, overtaking that for coal-fired power stations, and that there seemed no reason why this trend should not continue. The committee also expressed criticism of the secretive ways in which the country's energy policy has been determined and of the vulnerable and potentially troublesome links France has with Iran through the 'Euradif' enriched-uranium project.

With only weeks to go before a general election, the French Socialists came out for a nuclear moratorium. Their leader, François Mitterand, attacked a programme 'launched like a railway engine at 400 kilometres an hour', complained that decisions were taken in conditions of technocratic secrecy, and called for people to challenge a strategy which would lead to their 'being massively dependent on a nuclear programme whose assumptions are come daily more into question'. The Socialists pressed for an eighteen-month abeyance, in which to investigate energy-conservation measures and 'public inquiries on the British model'.

But the French Socialists are by no means united on the issue. A powerful lobby within their ranks is concerned, inevitably, to preserve jobs in the industry – where some 100,000 jobs are said to be at stake. According to some foreign observers workers within the nuclear industry in France are opposed to their employers' aims and objectives. But dispute, as in Britain, has

87

been confined to negotiating wages and conditions, not to discussing fundamental questions of the kind raised at the Windscale inquiry. There, the official submission from BNFL trade unions was an articulate but modestly phrased declaration of support for the company's plans and for present national energy policy as endorsed by the Trade Union Congress. Peter Adams, chairman of the BNFL joint industrial council – a federation of the seven unions represented on company sites – concluded his evidence to the inquiry by saying that 'there will be more deaths, more pollution in the atmosphere, more accidents as a result of the importation of Japanese motor cars than there will ever be with the reprocessing of their fuel'.

In Britain, general policy statements by the TUC are regarded as both determinants and reflections of Government strategy, especially during a long period of Labour administration: the feedback is positive and obvious. So the members of affiliated unions must feel that they are acting and speaking not just in their own local interests when they invoke TUC philosophy. The dissenting Yorkshire coalminers, represented at the Windscale inquiry by Arthur Scargill, were regarded with chilling hostility – bordering on hatred – by the BNFL workers. Scargill was accused of an ignorant betrayal of the labour cause, of special pleading for an industry which, by comparison with nuclear power, had an inordinately bad record for occupational health and safety.

In fact Scargill was out of step with most members of the group that sponsored his appearance at the inquiry – the Society for Environmental Improvement. While the SEI takes broadly a no-growth approach to energy generation and industrial development and especially resource-extensive technologies, Scargill's vision is of a steadily expanded consumer economy based on alternative fuel sources, of which solar collection is at the top of his list. Idiosyncratic to the last, he insisted that no sacrifice – even the loss of his own coalmining industry – would be too great for an immediate cessation of *all* nuclear activity. In this, he went much farther than most of the objectors to THORP, few of whom have called for or contemplated an early closedown of existing nuclear facilities.

But *their* arguments are no less strongly resented by nuclear-industry unions, who perceive much of the opposition to nuclear expansion as a middle-class assault on their employment prospects and their intelligence. Many environmentalists do not seem to have grasped the point firmly enough, whether in the nuclear industry or elsewhere: if livelihoods are involved, then they are inviting conflict and retrenchment unless they can make a positive case for alternative employment.

In Britain at least, there are possibly more objectors to present and proposed nuclear developments than there are defendants. But those in the second category cast far more economic and political clout. Vital as the issues may be, there are very few votes to be gained – and a lot to be lost – through any direct challenge a politician might make to the conventional economic wisdom. If anything, this means that the conflict remains essentially one between Government and/or industry and an increasingly militant and frustrated lobby which has no *real* political power. Unless there were a drastic and dramatic upheaval of social and economic priorities in the West as a whole, then an environmentalist mandate is most unlikely to emerge.

Internationally, the controversy is less tractable still, for it hinges heavily on the suspicion and distrust in which the industrialized world is widely held for its contemporary attitudes to development. The dilemma is compounded by the fact that, while Western statesmen begin to preach a gospel of caution and agencies like those within the UN promote soft-technology options to solve Third World problems, the principal commercial and political institutions of the industrialized world press their case for orthodox patterns of growth in the manufacturing and energy-supply sectors, whether they are dealing with Bangalore or Birmingham. These issues are bound up inextricably with the largest single global problem we face – the massive and growing inventory of arms, particularly nuclear weapons. It may be possible to solve the problems *and* preserve the overall peace – but not without extensive, open and painful confrontation, of which Windscale was just a quiet beginning.

7 Windscale: A Little History

The operation of the first generation of nuclear reactors and the reprocessing of their burned-up fuels, as described earlier, have become technological commonplaces. Windscale is not a particularly complex chemical plant. The fundamentals of separating irradiated nuclear fuel into re-usable fuel and waste have not changed all that much since the factory began operations, under the old Ministry of Supply, in 1951. Ownership and management was transferred, in 1955, to the UKAEA as defence requirements gave way to commercial ones. In 1971, BNFL was formed as a Government-owned limited company, providing a complete nuclear victualling service to the Central Electricity Generating Board and the South of Scotland Electricity Board, as well as contributing research and development services to the UKAEA, which holds the company's shares on behalf of the Government.

BNFL employs around 4,500 people at Windscale, of which two thirds are manual workers. Like their fellows elsewhere in the process industries, they perform for the most part repetitive and unexciting duties. If there is a significant difference it is that some Windscale workers – those required to work in active areas – are subject to the use of irksome special clothing and protection systems; in addition, the workforce is subject to the exceptional provisions of security legislation such as the Official Secrets Act and the Nuclear Installations Act. Though a few perceive the potential for erosion of normal industrial rights in this, most employees are no more and no less content to clock in and out of the Windscale plant than if it were a biscuit factory.

To that extent, they were surprised when their workplace became a word on the front pages of the national newspapers and in the broadcasts of radio and television. They were resentful

– and still are – of critical comment, however well-informed, about what they did at Windscale and had done for more than twenty years. Their work, until then, had been regarded as straightforward and uncontentious within the industry, and almost completely ignored by those outside it. The work was – is – routine. Fuel from power reactors is stored to cool, mechanically and chemically separated, and then processed, stored, redistributed. By the early 1970s, BNFL – confident, profitable and acknowledged as a world leader in its field – drew up plans for a major expansion and announced them publicly in late 1975: they included the proposals for THORP.

The environmentalist movement had, until 1973 or thereabouts, left the nuclear industry alone. Its major plants were geographically remote and its affairs – for most people – technically impenetrable. It was not a visibly polluting industry. And in Britain it had suffered only one serious accident, when the No. 1 reactor pile at Windscale caught fire and released radioactive contaminants over the area. A Government report (Cmnd 302, November 1957) on this accident does not fully reveal the disruption this fire caused. Although suitably grave announcements were made to the public on radio at the time, counselling the usual calm, the industry and the Government effectively went into a state of red-alert. The fire, poorly quenched with water, burnt while very large volumes of milk were contaminated. The chains of command for taking emergency action were ill-defined and, at some points, non-existent. In Cumbria County Council's report on the background to BNFL's application to construct THORP, the local borough council, Copeland, noted that 'very guarded comments were made on the "Windscale Incident"'. The fact is that many people were left uninformed. The fire was eventually put out, the reactor pile shut down and sealed, local milk consumption controlled for a while, and a business-as-usual posture adopted.

The accident was among the events and concerns that led to the formation of a Local Liaison Committee in the same year, charged with drawing up emergency plans and informing the local public about developments at Windscale. The committee, as Parker notes in the inquiry report, has plainly failed to meet its original brief. Its chairman has been automatically a representative of BNFL; the Committee has met on company premises; and minutes of its meetings have not been made public. Attempts by outside bodies to be represented on the LLC have been rebuffed: as late as 1976, when the THORP plan had been announced and general public controversy had begun, the local branch of Friends of the Earth – FoE West Cumbria – were refused information or co-operation from the Committee. Even formal bodies were still uncertain what was required of them in an emergency.

That will change when the recommendations made by the Flowers Commission and by Parker are implemented. That the recommendations should have to have been made, however, is a measure of the unsatisfactory state that prevailed and which arose for two reasons. One was the tradition of military state secrecy inherited by the UKAEA and, later, BNFL – characterized by concealment and a defensive response to questions. The other was a generally acquiescent frame of mind among local politicians and other opinion-leaders in this semi-rural part of the country. Another factor which undoubtedly has minimized the attention given to Windscale is that it lies just outside or only just inside the four television stations which serve the region – BBC North-West, BBC North-East, Border and Granada. Until the mid-seventies, it was forty miles from the nearest local radio station and from the regular correspondents for national newspapers. There is now a small studio in Whitehaven, serving Radio Carlisle.

The company's public-relations activities through the sixties and early seventies consisted mainly in arranging tours of the Windscale and Calder sites and of providing speakers for non-partisan organizations and school careers-advice sessions. Though a group calling itself Half-Life had begun campaigning in Morecambe and Lancaster (near the Heysham nuclear power station) against nuclear power, and the *Daily Mirror* had published its 'nuclear dustbin' feature on the Windscale expansion plan, the company was not prepared for the volume of press and public interest that had developed by early 1975.

From a technical point of view, it had no need to be too

worried. At the major press conference and tour of the plant called in late January of that year and held at Windscale, few of those invited were able or inclined to put hard questions. The question and answer session was monopolized by Walter Patterson, of FoE, and Anthony Tucker, science correspondent of the *Guardian*, both of whose questions might as well have been phrased in Finnish for all the understanding they generated among surrounding reporters. A Granada television director was busy setting up shots of the Calder Hall cooling towers, their steam rising into the winter air: it was, he told his reporter and crew, a good place for a filmed interview in front of the 'polluting Windscale chimneys'. For the most part, newsmen settled for a low-key 'row' story (BNFL *vs* environmentalists) and a generous buffet lunch.

BNFL was not, at that stage, practised in the more popular techniques of lobbying. It did not, as it does now, take the offensive. Its spokesmen on radio or television – usually senior company officials – invariably gave the impression of being hurt and surprised by the idea that anyone would want to challenge their views or their employer's policies. Their reaction to critical articles and letters in the national press was ponderous. Even now, the company spends more time rebutting the opposition's arguments than it does, in print, advancing its own. And only after a combination of Ministerial directive and sheer expediency has it altered its approach to the publication of information about accidents.

There are, on average, ten accidents or 'incidents' at Windscale every year. The great majority of these are minor, even trivial ones. But once the public interest in nuclear affairs had developed, each one was a potential source of news copy for the press and of embarrassment for the company. Whether apocryphal (as some were) or real, the effect on BNFL was the same. No more attention, for instance, was paid to the serious fire, explosion and blowback which put the first oxide-reprocessing plant out of operation (the subject of another Government report, Cmnd 5703, in July 1974) than was paid to dark and conspiratorial tales of radioactive waste spilled on roads and attended to by night-time decontamination squads. It would have been in BNFL's interests to make any accident or untoward event, whether serious or not, a matter of public knowledge, but the company pursued its policy of secrecy, of which – even in 1978 – it is not entirely cured: it took a chance discovery by a Border Television reporter to inform Carlisle City Council that the city's airport was being used to fly BNFL fuel cargoes between Dounreay and West Cumbria.

There was, and still is, the more conventional lobbying taking place on its usual two levels: the one which operates almost unconsciously through the institutional networks and which produces a consensus of industry-assumptions; and the one applied directly to and through Parliament, using the classic pressure-group tactics. A little late in the day, perhaps, these are now employed by the environmentalist movement, which has recognized that power is as likely to spring out of the barrel of a bar in the House of Commons as from anywhere else.

In 1975, as confrontation developed, the company began to spruce up its PR operations. A major move was the hiring of a Director of Public Information. Harold Bolter, a highly able and experienced industrial reporter from the *Financial Times*, was brought in to control the handling of news. Almost immediately, the company's persona began to change and become altogether more buoyant and aggressive. It still had a difficult passage ahead, but the new man, having spent many years winkling information out of or about reluctant companies, knew how to identify and repair weak spots in BNFL's guard. The policy was forged and polished over the next eighteen months and emerged gleaming at the time of the inquiry: in both the presentation of its case and the supply of background and clarifying material for the press, BNFL could not have seemed more agreeable, co-operative and open.

Any statement that might have been open to an unfavourable interpretation was spotted and an attempt made to correct it. Bolter, who takes one of the fastest and most accurate shorthand notes in the business, was frequently called upon to help pressmen at the inquiry who, through other commitments, laziness or lack of shorthand, had missed crucial exchanges. (I was one of the reporters who had cause to ask his help on all three counts.)

In addition, he was able to suggest to the company's legal team and their witnesses what points were likely to be taken up by reporters and thus which ones should be emphasized or softened.

The new responsiveness was highlighted on the evening of 22 March 1978, when the Parker report and its findings were being debated in the House of Commons. A report on BBC television's 'Look North' included an interview with Dr John Cunningham, Member for Whitehaven, Under Secretary for Energy, and a firm supporter of the THORP project. In the absence of an opposing point of view, it might have been better to use an independent commentator – a BBC Parliamentary staffman, perhaps. But such imbalances often occur for perfectly innocent logistic reasons. The item was double-headed, however, with one on the previous day's decision by Carlisle City Council to permit flights of nuclear cargo in and out of the city's airport, and the programme – which had a month earlier lost its Cumbria reporter, Jake Kelly, to work as a public relations officer at Windscale – ran a BNFL-produced film showing safety testing of the containers taken on these flights. After the film, which carried a voice-over script by the BBC's man in the Newcastle studio, was an interview with Peter Wilson, head of the Plutonium Business Centre at Risley, who explained that the containers were fool- and terrorist-proof. No other comment was included.

While the company prepared its departure from the old tradition, the public campaign against reprocessing was undergoing a similarly dramatic change of style. When the *Daily Mirror* had reported, in late 1975, that a contract was being negotiated for the reprocessing of Japanese spent nuclear fuel, a loosely concerted opposition began to assemble. Half-Life and the Conservation Society mounted a demonstration in the port of Barrow, protesting against the arrival of spent fuel from Sweden destined for Windscale. There was a flurry of critical correspondence in the newspapers; and there was a short House of Commons debate. The Energy Secretary, Tony Benn, suggested that there should be a public discussion of the Japanese contract, and BNFL convened one in Church House, Westminster, in

January 1976, when members of Half-Life and FoE met senior officials from BNFL for the first time.

The discussion was the first real one there had been between the respective adversaries, and, significantly, was attended by Benn himself, who talked of the opportunity for launching a 'great debate' on nuclear policy. How much of a debate it then was may be gauged by the fact that the Japanese negotiations were given Government blessing just two months later. The environmentalists redoubled their campaign and, a few weeks afterwards, organized a mass open-air teach-in by the perimeter fence at Windscale. It was attended by about 1,000 people, nearly half of whom had travelled there in a specially hired train – another classical British component of the lobby process.

Before a planning application for THORP had been formally submitted to Copeland Borough Council, in June 1976, the County Council had already decided that some form of prior public consultation was called for. It had received the first of many hundreds of letters and communications, including some early detailed submissions from formal lobby groups. Ordinarily, Copeland would have granted permission for the Windscale development, particularly in view of the employment prospects it seemed to hold. Only one member of the Borough Council, Bill Dixon, had reservations about the application. He was refused a 'conscience vote' when the application was later approved, and appeared at the inquiry as a witness for FoE West Cumbria. In December 1977 he was sacked as the Council's deputy chairman for the Labour group. The group's chairman explained that 'No one would be against his speaking to the inquiry . . . but he did not tell the group what he was going to say.'

The formal planning application, which then – to some observers' surprise – included details of less controversial work to be carried out in addition to the construction of THORP, was referred to the County Council, the Economic Development and Policy and Resources Committees of which were broadly and warmly in favour of granting permission. They had, how-ever, agreed that a public meeting of some kind should be held, and the Planning Committee Chairman, Stephen Murray – had urged the County to seek independent professional advice on nuclear matters. Professor John Fremlin, who holds the chair of applied radioactivity in the Department of Physics at the University of Birmingham, was appointed as a consultant. (His evidence to the inquiry suggested strongly that the risk of accident and untoward environmental impact from THORP would be tolerably low.)

Murray's contribution to the controversy will, in time, be seen to be a shrewd one. He was acutely aware that, though the county as a whole was strongly in favour of the BNFL plans, it would be seen as a peculiarly craven and ill-advised administration if it were to give an outright vote of approval. He urged on a public meeting in Whitehaven, which was held in September 1976 and attended by 800 people. The proceedings were televised and taped, were widely covered by the local and national press and were broadcast live by BBC Radio Carlisle. The meeting divided into three camps: those in positive favour; those against; and those who felt that – whatever was said there – the issues should be resolved in a much more formalized fashion at a full public inquiry. Murray had already prepared the way for this by pleading, in a letter to *The Times*, for the Secretary of State for the Environment, Peter Shore, to call in the application.

There was no response to this letter, and Murray recalled, in his evidence to the inquiry, that 'I did not think it appropriate, as Chairman of the Committee, to approach the Minister direct to seek him to call in the application. This, in my opinion, would have been an abdication of the duty of a local planning authority to determine a properly submitted planning application, and would have indicated a lack of confidence by the County Council in its capacity to determine the application . . .'

It was, and continued to be, an exquisite piece of political balancing. The County Council had received and considered a large number of submissions from objecting groups and individuals; it had awoken to the fact that many of the opposition arguments were not run-of-the-mill worries about pollution; and it knew that its position *per se* as a local authority – a large and reconstituted one following the reorganization of 1974 – was on trial. By October 1976, the debate had been joined by yet more groups – the National Council for Civil Liberties, the Council for the Protection of Rural England and the Town and Country Planning Association among them. The controversy had also been inflamed by publication of the Royal Commission on Environmental Pollution Sixth (Flowers) Report, which – though couched in vaguery – was critical of much of the nuclear industry's conventional wisdom.

At the beginning of November, Murray's committee agreed that, if the Environment Secretary was *not* going to call in the application, their best course would be to treat the application as one that departed fundamentally from the County's development plan. This was a doubly subtle strategy. It brought before Shore an in-tray problem he could not pass on; and it gave opponents the opportunity, if Shore did not treat the application as a departure, to proceed legally against the Council on the grounds that it *was*. As Murray noted, '. . . even if such litigation was ill-founded and unsuccessful, it would certainly have caused great, and undesirable, delay'.

Not everyone in the County was concerned with such subtleties. The Director of the Planning Department, Gordon Fanstone, clearly unused to organized protest in this part of England, was moved to write an astonishing letter to one of the local organizers of opposition to THORP. In late November he wrote to Edward Acland, a Kendal councillor and co-ordinator of the group Network for Nuclear Concern – who for some time had been whipping in support for the objectors – as follows:

It has been brought to my notice that you have been writing letters to the press calling on the County Council to act in some way or other in respect of some action you purport has taken place. It is also alleged that there is in existence a petition of some sort which has been gathered somewhere and delivered somewhere else with some unknown objective.

I have no record of the County Council having received any communication from you regarding this or many other assertions and statements which you appear to have released and I should have thought it would have been elementary courtesy to say nothing of correct procedure in relation to responsible government to contact in some way the body which is handling the [THORP] application.

The media in general and the press in particular are not the government of this country or of Cumbria and it is indeed peculiar, and unprecedented in relation to a formal planning application, to experience the techniques used by the individuals you claim to represent.

Is any information to be expected?

It was. In the next few weeks, British Nuclear Fuels and the County Council could hardly have had a worse press or the objectors a better one. Shore continued to play for time – twenty-one legally allowed days, to be precise – in which to consider the Council's formal turning-aside of the responsibility. Letters to the press and to MPs called for the convening of an inquiry. The House of Commons heard a motion calling for an inquiry. The Network for Nuclear Concern collected 27,000 signatures in Cumbria calling for an inquiry (and the Windscale workers collected 18,000 signatures *for* the project). Other organizations, some newly formed, others established, joined the campaign. On the eleventh hour of the twenty-first day, Shore announced that he needed time to reconsider the application (FoE had already asked that the BNFL proposals be resubmitted in three parts, in order to consider the THORP plan separately from the other, less contentious, parts).

The planning committee did in fact agree to accept an amendment to the original application by deleting all references to the proposed oxide-reprocessing plant and approving the remaining parts of the application. Four new applications were later submitted by BNFL and considered, at a meeting in Kendal, in April 1977. Before then, an inquiry had been decided upon, and was announced by Shore on 22 December 1976. He was responding, unquestionably, to the relentless pressure put on him by the environmentalist lobby. Set against that were equally persuasive forces being exerted by the Department of Trade and Industry, then preparing a capital-loan plan for BNFL's expansion, and anxious – in the very gloomy economic mood that then prevailed – to support anything that seemed to promise a return in foreign currency. Tony Benn, who had sanctioned the negotiations of Japanese spent-fuel reprocessing at Windscale, was just as painfully squeezed by his commitment to open government and industrial accountability. Recurrent accounts of 'incidents' at the BNFL plant did not fit in with his stated concern for participation and the democratic management of the working environment.

On 12 October 1976, BNFL discovered that a concrete storage silo was leaking up to 200 gallons a day of contaminated water into the surrounding soil. The leak, found during the construction of a storage extension for reactor wastes, could not be pinpointed (it has still not, in spite of gargantuan engineering attempts to identify and plug the seepage). The leak was not notified for twelve days, a delay which is generally thought to have incensed Benn, who thereafter directed that *any* abnormal occurrence in the industry should be immediately notified and made public. His rebuke to the industry is widely regarded as a decisive factor in calling for the Windscale inquiry. Close observers say that it was a tactical gift to the Energy Secretary anyway, who had long been unhappy about the nuclear industry's political performance and who now had an irrefutable case to prosecute for a full and open examination of its affairs and the issues raised by THORP.

Peter Shore did not refer to the silo leak, but he was quick to deny that he had been influenced by informal forces. He was asked, in the Commons, for assurance that he had not 'given in to the extravagant fears which seem to be gripping the country about nuclear energy' (Arthur Palmer, Labour MP for Bristol North-East). Shore replied that he had taken notice mainly of the Cumbria Council's planning report 'and not of the less authoritative voices which have been casting doubts on the development of nuclear energy'. Who those voices were it is not certain. Cumbria, after all, had declared its unwillingness to play the expert or the authority on nuclear matters; and among those who did express opposition were some eminently fitted to speak their mind, as letters in *The Times*, the *Guardian*, the science journals and international newspapers amply demonstrated over the following months.

BNFL and the UKAEA dropped their PR guard somewhat over the next few months, engaging in wounded correspondence, publicly and privately, with the objectors. In March, they resubmitted their application in four parts, three of which were additions or extensions to the Windscale pond-storage facilities. One of these applications was directly related to the building and operation of THORP, and the Council's planning committee were concerned that approval for this might prejudice the inquiry. But they were urged to give approval and told that BNFL would not be free to sign the Japanese contract unless it could store Japanese fuel in the proposed new ponds – where the fuel would remain in store until THORP was working. The committee decided to grant permission on condition that the fuel would be removed if planning permission for THORP was refused. (There is, as some have noted, no guarantee whatsoever that such a condition could be enforced: if THORP was a total technical failure and the Japanese or any other foreign government at some later stage decided that it did not want to retrieve its spent unreprocessed fuel, Britain would be left with it.)

On 25 March 1977, the application for the oxide-reprocessing plant was formally called in and a date set for the inquiry – 14 June. The environmentalists were thrown into a state half way between delight and distress. What should they now concentrate on: preparing their case, or raising the very large funds necessary to pay for it? They had to compromise and do both at the same time. Two of the major groups fighting THORP – the Isle of Man Local Government Board, and, to a lesser extent, the Town and Country Planning Association – were not too worried about finance. The other two – Friends of the Earth, and the Windscale Appeal Group, an *ad hoc* federation of nine environmentalist and conservationist bodies – were desperately short of money but contrived, through newspaper advertisements, appeals and fund-raising events, to pay their initial costs of retaining and briefing counsel, and of covering administrative, research and travelling expenses. For a great many of these and the other, legally unrepresented, witnesses opposing THORP, a lot of the cost was to be borne out of individuals' own pockets.

Not surprisingly, there was disagreement between groups about how their resources should best be divided. During the course of the inquiry, some felt that funds raised or underwritten for collective use had not been as effectively allocated and deployed as they might have been. Looking back on the hearings, however, I think it unlikely that the force and style in which respective cases were put would have been significantly different if the objectors had had much longer to prepare their cases, consulted one another as to how the issues should be jointly or independently tackled, or spent more money in the assembly and presentation of their evidence. That is not to deny their criticisms outlined in Chapter 11 – of the imbalance of resources available to the principal consultants.

8 Windscale: Inquiry Diary

Within a week of the call-in for a public inquiry of BNFL's application, an Inspector and two technical assessors were appointed. The Tribunal consisted of the Hon. Justice Parker, a High Court Judge, who, as Roger Parker, QC, had been a barrister specializing in commercial cases; Sir Edward Pochin, CBE, MD, FRCP, head of the Department of Clinical Research at University College Hospital, London; and Sir Frederick Warner, DSc, CEng, FIMechE, FRS, senior partner in the engineering-consultancy firm of Cremer and Warner. Parker had chaired the inquiry into the Nypro chemical-plant explosion at Flixborough in 1974. Pochin is a former chairman of the International Commission on Radiological Protection and a member of the National Radiological Protection Board, for whose chairman he has deputized. Warner was a member of the Flowers Commission, which produced the Sixth Report for the standing Royal Commission on Environmental Pollution – 'Nuclear Power and the Environment': his consultancy has carried out work for the nuclear industry.

On 17 May 1977, a few weeks before the inquiry opened, Parker held a preliminary meeting in Whitehaven, at which he suggested procedures to be adopted at the hearings and also made a number of remarks calculated to cheer many who, otherwise, were ready to believe that the inquiry was to be a whitewashing exercise. He noted, first, that although the inquiry was a 'public local inquiry' under the Planning Acts and thus in no way different from an inquiry into *any* planning proposal, it was nevertheless unique in other ways: '. . . the issues to be investigated may affect not only those already alive and residing in the immediate neighbourhood but also those who live far away and who will not be born for many years ahead.'

The scope of inquiry was thus to be very wide. Among the implications to be questioned would be those arising out of transport and storage of spent fuel, environmental hazards and occupational risks in the operation of the proposed plant, security risks and industrial relations. Parker had been supplied with press cuttings in some of which he had seen 'expressions of disquiet lest the inquiry would be unduly limited in scope. I trust that what I have said will allay any such disquiet.' The inquiry was to be unique in that, for the first time, the arguments and evidence of both opponents and supporters of THORP were to be tested by cross-examination in public. He quoted an American study (the Ford Foundation report on nuclear power: 'Issues and Choices') which suggested that the nuclear debate was generally undisciplined and poorly structured. Parker was concerned to see that no lack of discipline and structure would characterize *his* inquiry:

'[The Flowers Commission] say "We have noticed that the debate is not always well informed, that sometimes relatively minor matters receive attention to the exclusion of others potentially more important, and that the context is often poorly defined." I have no doubt that there will be dedicated advocates of various views appearing at the inquiry to present arguments and give evidence as witnesses. Because the issues are grave and because they arouse strong feelings, the temptation to resort to polemics may well be great, but I hope that it will be resisted. The polemics have served their purpose.'

His task, he said, was only to investigate, report and recommend: this was the only way in which the public interest might be advanced: 'I will, for my part, do my best to assist by keeping a largely flexible programme and by making sure that evidence is *thoroughly tested*. I have also made arrangements to ensure that, in so far as my assessors are unable on the spot to check the accuracy or validity of calculations, formulae, and the like, they may be checked by independent sources' (emphasis added). There were to be two significant departures from normal inquiry procedures. The first was that opening statements might be made by all parties, not merely by the applicant: where this was not done, Parker observed, 'the opponents' cases sometimes tend to get less than equal treatment in the press and on radio and television'. Similarly, closing speeches would be admitted from the objectors as to the final effect of all that would have been heard.

The media were to be provided with all the necessary facilities for covering the inquiry (generous and helpful ones, as it turned out), but sound and film coverage of the hearings was not to be allowed – so that 'no party or witness should feel in the slightest way inhibited or, on the other hand, tempted to speak past me or his opponents to the crowd'. Lastly, he was not prepared to admit evidence which would be contrary to the national interest or which would call into question the merits of any Government policy. Neither of these was more than vaguely defined: the proper place to debate Government policy, Parker asserted, was in Parliament, though this should not inhibit argument about particular recommendations that might or might not be included in his report. He could not, he said, allow evidence that might prejudice national security or assist terrorists and others to gain information about nuclear weapons and materials.

For the most part, objectors went away from this meeting somewhat mollified, a few going so far as to express aloud their satisfaction at the choice of Parker as Inspector. When the inquiry opened a few weeks later, on 14 June, their feelings were reinforced by his brisk, attentive and apparently impartial treatment of all concerned. Few of those present had ever watched or listened to a judge presiding over the submissions of senior counsel. Parker's direction that all evidence be entered and examined under oath – a rare procedure for a planning inquiry – added to the strangeness and seriousness of the occasion, though the seriousness was naturally offset by the usual courtroom banter (which, if anything, serves to further distance ordinary people from the law). This excerpt from the daily transcript is typical:

THE INSPECTOR: I think I have a missing page. That is all that worries me.

MR KIDWELL [*for FoE*]: It is one, two, and three. May I say I do not want page two and three.

THE INSPECTOR: You only want page one?

MR KIDWELL: Yes.

THE WITNESS [*Conningsby Allday, managing director of BNFL*]: We should have the complete letter, should we not?

MR KIDWELL: You should have one, Mr Allday. You have not got it?

THE WITNESS: No.

THE INSPECTOR: Does page two end 'Far from having the effect of increasing electricity prices to pensioners . . .'?

MR KIDWELL: Yes.

THE INSPECTOR: And page three goes on '. . . of about £5m.'?

MR KIDWELL: No.

THE WITNESS: No. That is Sir John Hill's letter.

THE INSPECTOR: I have not got page three. I had better have it.

MR KIDWELL: Yes, that appears to be the first page of another letter.

THE INSPECTOR: It might be as well to have it assembled overnight in the proper order. In the meantime, you might possibly tell me how much longer you expect to take with Mr Allday?

MR KIDWELL: I would think, sir, that I would be finished by lunchtime tomorrow.

THE INSPECTOR: Thank you. Lord Silsoe, you are looking almost pregnant; have you got some point you wish to make?

LORD SILSOE: Absolutely none at all, sir. I just look like that. (*Laughter*)

[*The Inquiry Was Adjourned Until Ten O'clock The Following Morning*]

It was not long before some objectors, particularly those without legal counsel and who were paying their own expenses, largely for research, travelling and accommodation costs, were beginning to complain among themselves that such clubby frivolity had no place in the hearings – especially when, as they believed, the Inspector was applying sterner criteria to them than to the heavyweight counsel and their witnesses. Even legally represented groups, however, were worried by the prospect unfolding before them of a very long inquiry which would exhaust their financial and physical resources. Some had already submitted that they had had too little time to prepare their case and raise the necessary funds – a submission wholly rejected in the Parker report. In the event, funds *were* somehow found, arrangements arrived at between objectors and their lawyers about payment, and a gradual adaptation made to the fact that the hearings would last much longer than anyone had previously

expected. In many cases, the objectors were able to prepare their
evidence as they went along, practising their ability to cross-
examine, revising their strategies and recasting their own evidence as they heard what the applicants had to say over the first
ten weeks or so. They could also reappraise the priority of their
objections, which individually and collectively shifted a good
deal over the course of the inquiry.

The inquiry was set in Whitehaven Civic Hall, a light, airy,
better-than-average modern piece of municipal architecture
close to the docks of this West Cumbria port. Objectors stayed in
lent or rented houses, caravans and even in their own cars and
vans, where they discussed and prepared many of their submissions. BNFL maintained a large research and legal-support
staff in a near-by school. Reporters took over a congenial pub in
the little seaside village of St Bees. Parker and his assessors lived
in Tallentire Hall, a parkland-shrouded, security-guarded stately
home near Cockermouth, a thirty-mile drive away by chauffeured
Austin 1800 (the Judge occasionally used his own elderly
Alvis).

For the first few days of the inquiry, the hall was crowded with
local people and others who had come to watch the proceedings.
After a week, though, there were seldom more than twenty non-
participants in the main hall; a second, adjoining room set aside
as an overflow hall, was used mainly by objectors to brief their
counsel and revise evidence, and by journalists, who could
listen to the hearings on a relayed public-address system: the
40,000 or so words spoken each day were taped and the tapes
erased after transcripts had been typed. These were copied
and circulated, for a token £1 per sitting, usually on the following
morning: invariably impeccable in clarity and accuracy, the
transcripts were prepared by a team of professional shorthand
writers from Sheffield and distributed by a Department of
Environment/Central Office of Information team.

The work of this team epitomized the departure from normal
inquiry practice. It co-ordinated the activities of the Tribunal's
own secretariat, the legal counsel, witnesses, press and general
public; planned the timetable of appearances – a major logistical
exercise in itself and subject to continual revision; and operated a

system to file, index and distribute the 1,500 separate documents entered as evidence. For veterans of planning inquiries, it marked a new mood of helpfulness and (relative) accessibility of information. Some of the objectors also managed to obtain direct help from the BNFL secretariat: that may or may not have been as much a matter of shrewd public relations as anything else, but it was still valuable and value-free technical assistance.

Throughout, the two officers from the Special Branch sat in on the hearings, looking either bemused or bored.

At an early stage, there were discussions, both informal and in submission to the inquiry, as to whether this was an appropriate and satisfactory way in which to resolve the application at issue. The Isle of Man Local Government Board suggested, through its counsel George Dobry, QC, that there were, in any event, procedural defects in the application itself or in the steps leading up to the inquiry. This submission was later dropped, but another – that an environmental impact analysis (EIA) along lines laid in US planning inquiries be substituted at least in part for the adversarial techniques employed at Whitehaven – was advanced. This idea, presented by Sir Frank Layfield, QC for the Town and Country Planning Association, is examined in Chapter 11.

What follows is a very compressed account of the inquiry (daily and weekly coverage of the hearings appeared, in Britain, in the *Guardian* and the *New Scientist*; discontinuous coverage was given by *The Times*, *Financial Times*, and the radio and television networks).

Week 1 BNFL summarized its case for the construction of a thermal-oxide reprocessing plant based on

the need for increased separation and safer storage of irradiated wastes;

the need for a higher recycling capacity of uranium fuels for UK reactors;

the need to recover plutonium for possible use in fast-breeder reactors;

the potential market for reprocessing additional fuel from abroad.

The objectors made their opening statements. Broadly, they were concerned that the proposed plant would be a focus for voluminous international trade in nuclear fuels and wastes and would lead to infractions of the nuclear Non-Proliferation Treaty; they questioned any move to open the options for an energy economy based on FBRs; they feared the implications for security in a plutonium economy; they doubted that the sums had been satisfactorily done on energy economics; and they opposed the application on technical and environmental-hazard grounds.

The Isle of Man submitted that the concentration of plutonium in the Irish Sea around the Windscale effluent pipe was twenty-six times higher than that in the waters surrounding the Pacific island of Eniwetok used in the early days of atomic-bomb testing and evacuated for more than twenty-five years.

The Lancashire and Western Joint Sea Fisheries Committee said it foresaw circumstances 'in which the interests of the fishing industry could be significantly damaged' at present or higher rates of radioactive discharge from Windscale.

Week 2 Conningsby Allday, managing director of BNFL, continued his evidence for the company and faced cross-examination from counsel for the objectors and from individual opponents of the THORP plan. He thought a pause in the development of FBRs was inadvisable, and he gave an assurance that the plant could be safely shut down in the event of a mass walk-out. He did not concede the view – expressed by the Friends of the Earth – that uranium could well become cheaper and more abundantly available, which would damage BNFL's fuel-conservation case.

Outline details of the proposed contract between BNFL and the Japanese electric-power utilities were made available after informal talks between the company and some of the objectors. In return for taking 1,600 tonnes of spent fuel, BNFL would receive from the Japanese in advance more than a quarter of the cost of constructing THORP (estimated to be nearly £600 million). The customer would bear all technological and operational risks, and repayment liability would be limited to the

failure of BNFL to build the plant at all. Britain, the contract
makes clear, could be stuck with the recovered wastes if the
supplying country declined to take them back.

Week 3 B. F. Warner, deputy head of R&D at BNFL, described in detail the proposed reprocessing plant. By comparison
with the present Magnox reprocessing plant, THORP would
deal with between seven and ten times the fission-product activity, five times the quantity of plutonium, and thirty to forty
times more of the highly active transuranic isotopes. He also
recounted the causes and circumstances of the 1973 accident in
the oxide-reprocessing plant, when there was a release of radioactive ruthenium: his evidence referred to design modifications
which, in any future plant, would prevent the occurrence of a
similar accident.

George Dobry, QC, suggested that BNFL's application was
faulty in law. His objections, though technically valid, did not
derail the inquiry, as some present feared, but were entered into
the inquiry record. BNFL and FoE argued, somewhat fruitlessly, about the employment potential of THORP.

Week 4 The company presented its overall plans for nuclear-
waste management, and gave details of the HARVEST process
of vitrifying the most toxic effluents in borosilicate blocks and
'disposing' of them in geological strata. FoE challenged BNFL
to show why the reprocessing-vitrification route was an environmentally superior policy to that of storing unreprocessed wastes –
especially since plutonium was produced in the route preferred
by BNFL.

Peter Taylor, for the Oxford Political Ecology Research
Group, declared that one of BNFL's submissions was not a
fact but an attempt to gain credibility. Justice Parker, in rebuke,
rejoined: 'Do not get the idea, Mr Taylor, that everything that
appears in the proof is regarded by me as being evidence. It is
not.' To which Taylor, unbowed, answered: 'With all respect,
there are many people here who do not have your abilities to
follow exactly what is going on, and BNFL's credibility does
not rest just with the Tribunal. It also rests with the public.'

Week 5 John Donoghue, manager of the safety-assessment group at Windscale, described a computer routine, written for the company by the UKAEA's Safety and Reliability Directorate, to calculate the consequences of releasing radioactive material to the atmosphere. Data input for the program (known as TIRION) included those for quantity and type of material released, stack heights, weather conditions and population distribution. Playback, it was claimed, showed that the 'maximum credible release' would necessitate only a small number of evacuations from the area and a limited quarantine on locally produced food. Objectors were not impressed, and the Oxford group, PERG, asked that they be given the source program for TIRION to determine whether it really did simulate all the essential parameters.

Philosopher Karl Popper was quoted on risk by the Windscale Appeal Group: 'everything that is not impossible [for people to do incorrectly] is inevitable ...'

Week 6 Dr Brian Wynne, spokesman for the Cumbria-based group Network for Nuclear Concern, drew from BNFL the admission that the levels of radioactive ruthenium in edible seaweed (traditionally used in laverbread) suggest pathways in which the radiotoxicity has already exceeded maxima laid down by the ICRP.

PERG, having looked at the TIRION accident-simulation program, declared that it was the most accurate and sophisticated program available and 'a credit' to the Safety and Reliability Directorate. Nevertheless, it had severe limitations: it did not, for instance, take into account the possibility of a boiling-to-dryness of highly-active-waste storage tanks, did not allow for the possible violent rupture of cooling-pond walls at Windscale in the event, say, of an aircraft's falling out of the sky, and did not fully comprehend meteorological conditions. The group wanted to write an alternative program and lay a comparative analysis before the Tribunal.

BNFL described the precautions taken to ensure the safe transport of nuclear materials. This gave some technical reassurance about the engineering standards to which containers and specially equipped vehicles were built, but revealed that the responsibility for observing appropriate safeguards – as in many other industries – is spread among a large number of organizations and individuals whose work may not always be properly co-ordinated or supervised.

Week 7 Justice Parker received written answers from BNFL in response to his questions, a month earlier, about the production of plutonium at Windscale and the related implications for a possible programme of fast-breeder reactors in the UK. They showed that 7·5 tonnes of plutonium had already been recovered from reprocessed Magnox fuel and that about 45 tonnes would have accumulated from British fuel by the end of the century: of that, rather more than half would be required to fuel the first commercial fast-breeder reactor. A *second* reprocessing plant the size of THORP would be needed to sustain a full-scale CFR programme beyond the first decade of the twenty-first century.

FoE and the Town and Country Planning Association both argued that reprocessing would jeopardize attempts to arrest the proliferation of nuclear weapons. Dr Donald Avery, deputy managing director of BNFL, conceded, in cross-examination, that it was 'fair comment' to say that financial considerations had outweighed the dangers of proliferation, but went on to refute the idea that Britain, by going ahead with a reprocessing plant for oxide fuels, would be making a moral error: 'I am cynic enough to question whether self-denying examples do much good.'

Professor John Fremlin, appointed by Cumbria County Council to help resolve their indecisions about THORP, told the Tribunal that he had eaten fish caught off the coast near Windscale and had had his gastrointestinal tract monitored for radiopathogens. Fremlin, holder of the chair of applied radioactivity at the University of Birmingham, reported that his body levels of radioactive caesium after consuming the fish gave no cause for concern.

Week 8 Arthur Scargill, president of the Yorkshire branch of the National Union of Mineworkers, arrived at the inquiry amidst consternation from local union representatives – who support the BNFL proposals and who found his antinuclear pronouncements singularly lacking in both sense and solidarity. He was saying, they suggested, that they should face unemployment while the coalminers he represented went from strength to strength. In any case, they pointed out at a noisy impromptu press conference outside the Civic Hall, the mining industry had an appalling safety and health record, whereas the nuclear industry had a good one.

Scargill declared that he was not there to plead a special case for the colliers. In the long term, he believed that even coal would no longer be used and that solar and other benign sources could provide all the energy requirements of the UK. He was followed by a squad of witnesses for the Society of Environmental Improvement, who put forward detailed alternative-energy schemes – biological systems for synthesizing fuel, wave- and wind-power projects, tidal barrages and conservation measures.

Dr Vaughan T. Bowen, a geochemist and senior scientist at the Woods Hole oceanographic institute in Massachusetts, appeared briefly at the inquiry on behalf of the Isle of Man, and proceeded to lambast the Fisheries Radiobiological Laboratory for producing data that was in parts 'utterly uninformative'. A supporter of nuclear power in general (he was a chemist on the Manhattan A-bomb project), he nevertheless thought the Windscale site to be 'ill chosen' for a reprocessing plant.

Week 9 Bowen's allegations were refuted by the director of the FRL, Dr Neil Mitchell. He was asked by Parker whether some aspects of the Windscale operation which might otherwise have given cause for public alarm had been glossed over or presented in a favourable light by the Laboratory. No, he said: the bias, if any, was in the other direction.

The electricity Boards warned, in evidence, that any diminution or withdrawal of nuclear-fuelling facilities would seriously impair their ability to maintain power supplies. One Board witness agreed that it might be feasible to store unreprocessed spent fuels in cooling ponds but said that their fear was that serious fuel-deterioration would result.

It emerged that there was no positive plan for an inquiry into whether the UK would proceed with a programme of FBRs. The Department of Energy representative could say only that the question would be extensively debated 'in one form or another' at some future date.

Week 10 BNFL and the Central Electricity Generating Board produced – in response to an earlier request by FoE – a document that was to become central to the Windscale debate. In it, the company and the Board examined the costs of alternative fuel-handling techniques.

These showed that there could be a saving of £300–400 million if reprocessing were not adopted as a preferred fuel-cycling route, but BNFL insisted that a vitrification process would be necessary, whatever was done with spent fuel, and that this would make the storage of unreprocessed fuels quite uneconomic.

Justice Parker ordered a monitoring of plutonium levels in an estuary close to Windscale. The objectors maintained that a short-term sampling of aerial radioactivity could not be meaningful.

Week 11 Reg Farmer, safety adviser to the UKAEA, told the inquiry that an accident involving a chlorine road tanker could quite easily result in 200–800 'fairly immediate' deaths. A radio-active-transport accident, he submitted, should be assessed as a hundred times less hazardous.

Raymond Kidwell, QC, for FoE, referred to the BNFL/CEGB cost analysis presented to the Tribunal during the previous week. It restated, he claimed, what FoE already felt to be a 'lost position': to judge from the analysis, £300–400 million of British taxpayers' money was going to be 'literally reduced to sludge'.

The director of the Uranium Institute, Terence Price, argued that uranium finds would become fewer and more costly – supporting BNFL's fuel-conservation case for building THORP: the long-term integrity of fuel supplies could be assured only through a reprocessing–FBR cycle.

Week 12 It was announced that there *would* be a public inquiry into FBRs if the Government decided it wanted to go ahead with building the first commercial station (CFR1). The Environment Secretary, Peter Shore, made it plain that such an inquiry would be as wide-ranging as that into the reprocessing plant, enabling 'wider relevant issues' to be considered.

Dr Peter Chapman, director of the Open University's Energy Research Group, gave detailed and lengthy evidence on likely energy strategies for the future. In answer to close questioning from Parker and BNFL, he forecast that the most economical fuels within this century would be those derived from coal and solar sources; that appliance-saturation (and thus the heaviest potential electrical load) would have peaked well before the 1990s; that price-projections militated against the uranium–thermal reactor–reprocessing–FBR cycle; and that a major energy user – the transport sector – would transfer from petrol to battery-electrics. Work on high-energy-density batteries was promising: the forecast top speeds of 120 km and ranges of 300–400 km were well within the average daily requirements of most road users.

A net surplus in Britain's balance of payments – effecting the devaluation of overseas nuclear customers' curriencies – could, according to FoE, wipe out any profit in the reprocessing of foreign spent fuels.

Week 13 Professor Albert Wohlstetter, a leading witness for FoE, argued persuasively against reprocessing on the grounds that it would further enfeeble non-proliferation efforts. BNFL had said that reactor-grade plutonium was an inappropriate and unlikely ingredient for a bomb: this, Wohlstetter said, was irrelevant. 'Reactor-grade' simply meant that the plutonium had been lengthily irradiated and had a high proportion of the Pu_{240} and Pu_{242} isotopes. Light-water reactors characteristically operated irregularly and not in accordance with theoretical norms. They frequently discharged fuel rich in Pu_{239}, which could be used in weapons.

Wohlstetter also declared that initiatives on the use of nuclear power 'for peaceful purposes' were invariably a cover for activities with military implications. Atoms-for-peace projects were likely to be useful exclusively for war – and he cited Plowshare as an example.

BNFL reported that, between 1970 and 1974, there were forty-five 'incidents' at the Windscale plant; in 1975/6 there were fifty-four, nearly a third of which were the result of 'personnel error'.

Members of the local, West Cumbria, branch of FoE appeared before the Tribunal and argued a largely moral case against the proposed reprocessing plant. Among them was a member of the BNFL workforce, David Bainbridge, who told the inquiry that, although he and his fellow workers were familiar with radiation and its associated hazards, many felt that the plant and an expansion of it were not totally without risk. Another local objector, schoolteacher Andrew Dudman, showed by quotation that the supporters of the THORP plan had seriously contradicted their position throughout the hearings: it all illustrated 'the tendency for high technology to develop without the benefits of rational judgement'.

Week 14 Arthur Scargill reappeared to be cross-examined on his earlier evidence and was as bluff and buoyant as ever. Unmoved by a reminder that the Trades Union Congress and his own union's national executive were in favour of developing nuclear power, he declared that if only someone could tell him that there *would* be ('not "may be" ') no contamination, no technical problems, no nuclear accidents, no waste-disposal problems, no increased prospect of nuclear war and no problems arising from the production of plutonium, 'then of course I may view the matter differently'.

While Scargill was at the hearings, the National Coal Board was making test drillings for coal in the Whitehaven area (where only one working mine remains from the dozens that once dotted the littoral). Scargill commented that a reopened pit plus two fresh workings in West Cumbria would provide 3,000 new jobs – twice as many as promised from the THORP project.

Week 15 Quis custodiet ipsos custodes? Dr Brian Wynne suggested that there should be a more open and public system for the setting and keeping of radiological standards. Parker seemed unimpressed: he implied that recourse to Her Majesty's Stationery Office for information and the ballot box for political action together constituted an adequate mechanism for protecting public interest. He nevertheless asked Wynne to produce his proposals for improving the system.

Professor Graham Ashworth, head of urban environmental studies at the University of Salford, questioned whether West Cumbria was really an appropriate site for expanded nuclear-fuel-reprocessing facilities. He was concerned that any increased dependence in the area on nuclear-based industry could inflict suffering if nuclear energy were abandoned. The diversity of industry, he contended, should be as great as possible.

There was speculation that Parker might recommend, in his report, that Britain should have a 'nuclear ombudsman'.

Week 16 Sir Frank Layfield, QC, opened the case for the Town and Country Planning Association and raised the question whether it would not have been better to use an environmental impact analysis before making decisions on the proposed plant. Was he saying, Parker asked, 'that this inquiry is an inadequate means of determining the issues which have to be decided?' No, said Sir Frank: what he was putting forward was that there were a number of interrelated, key areas, in which the information so far available was not sufficient to resolve the THORP question. Without such information, it was not possible to make a satisfactory estimate of all the relevant risks – economic, technical, political, environmental – of going ahead with the project.

The TCPA put up several eminent witnesses, of which three lent particular weight to the Association's case: Professor Joseph Rotblat (non-proliferation); Dr Alice Stewart (epidemiology); and Professor Peter Odell (energy analysis). Dr Stewart's evidence was based largely on the Hanford study in the USA (conducted by Dr Stewart and Dr Thomas Mancuso), the tentative conclusions of which indicated that risks to workers in the Windscale plant might be twenty times greater than had previously been estimated.

Week 17 Still with TCPA, the inquiry heard more about EIAs from the Association's director, David Hall. An EIA, he said, could be happily fitted into Britain's planning system, and – if properly designed – would have given the Tribunal less, and not more, work to do.

BNFL told the hearings that, if permission was granted to construct THORP, it would contribute around £2 million towards the cost of local infrastructural support – land purchase, road improvements, house construction, training schemes and scholarships for students.

The Socialist Environment and Resources Association sees the development of nuclear power in general and the expansion of reprocessing facilities in particular as steps on the road to deskilling society and exacerbating structural unemployment. Dr David Elliott, a SERA witness, told the inquiry that it was appropriate to automate some unskilled and repetitive jobs, but the tendency was now to eliminate skilled and interesting jobs. Dr Elliott, a technology lecturer with the Open University, noted that the Lucas Aerospace Combine Shop Stewards Committee (which is confronting large numbers of redundancies) had outlined a number of projects on which its members could and should be engaged – the design and manufacture of resource-conserving products. In the USA, he said, it had been calculated that a solar-power programme could create two and a half times as many jobs as could a nuclear programme. The AFL–CIO had talked of reopening and retooling factories for a national conservation initiative; in Britain, AUEW–TASS at AEI Trafford Park had looked favourably on the possibilities for wind-power, wave-power and low-impact energy-storage systems.

Week 18 Another member of the Open University, Dr Barry Shorthouse, challenged the wisdom of going ahead with a full-size reprocessing plant on the basis of 1 : 5,000 scale pilot designs. BNFL saw what they said was a 'lack of logic' in his submission, but privately wondered why he had not approached them in his professional capacity rather than as an adversary appearing for the Windscale Appeal Group.

WAG witnesses also included Edward Goldsmith, editor of

the *Ecologist*, Dr Charles Wakstein, nuclear engineer turned film-maker, and Dr Kit Pedler, originator of the television series *Doomwatch*. Pedler had been ready to give evidence showing how easy it was for amateurs to construct an explosive nuclear device, using published materials and simple workshop technologies. Justice Parker thought it better for the evidence not to be read in public, and ruled that it be entered only as a document.

The Society for Environmental Improvement announced the results of an opinion poll conducted among 1,600 people in sixteen British towns and villages: they implied that large-scale public protest would follow a Government decision to build FBRs. The questionnaire used was, SEI admitted, a biased one. The Society's founder-chairman, Gerard Morgan-Greville, said that it was 'a way of forming public opinion as well as testing it'.

Week 19 Peter Taylor of PERG argued that, in all of the countries with which Britain has, or proposes to have, commerce in uranium, spent fuel and reprocessed materials, there is either militant antinuclear feeling or political instability. PERG, he submitted, would want to see a comprehensive and independent survey of attitudes in Britain before any decisions on THORP were taken.

Dr Brian Wynne, who also acts as an adviser to PERG, was recalled on his supplementary evidence. He produced a set of alternatives to what he had criticized as faulty institutional arrangements for decision-making and monitoring of radiological safeguards: he provided the Tribunal with a suggested 'family tree' of organizations within which environmental standards could be researched, set, authorized and supervised. He pleaded, again, for the Government funding of environmentalists' research, which he noted was common in Denmark, and for the replacement of the ICRP by a UN-administered body.

BNFL sharply challenged Dr Alice Stewart on her Hanford survey. She should, the company's counsel said, be prepared to recognize that the American studies on which her evidence had heavily relied suffered from 'grave deficiencies'.

Week 20 The week was dominated by the closing submission for Friends of the Earth, delivered by Raymond Kidwell, QC. In what was widely regarded as a brilliant summing-up for his clients, he argued at length that the nuclear Non-Proliferation Treaty could not be the safeguard on which BNFL and its supporters placed their reliance. The NPT, he agreed, was perhaps in urgent need of improvement, but the task of revising it would take much too long and could well prove to be fruitless.

Neither, he suggested, were sanctions a simple answer: the big stick had to be associated with a moral carrot whereby the British would say ' "We will refrain from producing a plant which is no use to us anyway" . . . [a plant] which is directly contrary to the spirit of the NPT, putting the raw material for nuclear weapons into the hands of non-nuclear weapon states. That cannot, in this year of 1977, be right.'

The great debate on nuclear power – which Kidwell had said at the outset was only now beginning – had taken place at Whitehaven in the best possible circumstances. Deferral of the THORP plant was, technically, saying 'No' – but in reality it was saying only 'No, not yet.' That required more courage than the ostensibly more forthright and outright pronouncement of a decision for or against the plant.

Week 21 The end of the inquiry. Lord Silsoe, QC, rose to make a long and meticulous speech for BNFL, rebutting the hundreds of points on which objectors had rested their case against the construction of THORP. The new plant, he said, was essential if the company was to honour its commitments to reprocess the 3,300 tonnes of spent fuel from AGRs and foreign thermal reactors which had already been contracted. The company had to deal with that fuel in a reliable and realistic way: reprocessing and vitrification would prove to be safer than the alternatives and would substantially improve the efficiency with which uranium was used.

There had, he went on, been no evidence to show that the need for nuclear power was in doubt, and this need could be satisfied in the long term by recourse to the fast reactor. Without

THORP, the UK might be subject to resource-blackmail if the country wanted to run a fission-power programme.

He produced a piece of supplementary evidence that suggested Dr Chapman's cost-analysis of the THORP project was based on an overestimate – by a factor of five – of unit reprocessing costs. Though the point was made with Lord Silsoe's characteristic diffidence, it seemed that a part of FoE's case had been devastated. He went on to say that, with THORP, Britain could make that much more of a contribution to INFCE (the International Nuclear Fuel Cycle Evaluation programme).

To suggest, as FoE had done, that Britain had no moral duty to help the Japanese economy, for instance (by reprocessing that country's irradiated fuel), would mean 'taking a large stride in the direction of a breakdown of international co-operation'. Again, there was no easy answer to the problems of proliferation and terrorism, but they would not go away even if terrorism did.

Iain Glidewell, QC, for Cumbria County Council, while giving broad support for the BNFL plan, submitted that there was indeed a need for improved institutional arrangements for environmental discharges from Windscale. Some steps 'over and above those recommended by the Royal Commission and the White Paper' were desirable. There should be greater dissemination of information at a local level, and a wider involvement in the issues at a national level.

The inquiry closed on 4 November, precisely a hundred days after its first sitting.

9 Parker's Justice . . .

The Parker report is one of the most important documents to be published in the past decade. In a few score pages, it demonstrates at one and the same time the political impact that the environmentalist movement has made and the failure of that movement to translate and transmit its real concerns. The integrity and, indeed, intelligence of Justice Parker the person are beyond question: his conduct of the inquiry reflected his undoubted skills as a professional lawyer and reinforced his reputation – earned as a top commercial silk – for the fine interrogation of formal evidence. What is at issue since the publication of his Windscale inquiry report is the role of Parker as an instrument of policy-making – as, if you like, a political receptor.

During the hearings, he showed himself – quite honestly and openly – to be ignorant of much of the mainstream environmental dialectic. That dialectic is not new, though public synthesis and recognition of it might be. It draws together the work of economists, anthropologists, statisticians, life-scientists and others over more than a century. It has been a live discipline for at least twenty years, during which time there has been a fusion of interests and expertise. Subsectionally, it is highly specialized – ranging from plant genetics to ekistics, agronomy to energy strategy, wildlife protection to resource conservation, population theory to pollution abatement. In sum, though, it has a common and comprehensible theme: a call for mankind in his use of technology to respect and protect the complex and interrelated global ecology; for industry to be more conserving of finite resources; and for it to be more closely fitted to the fairly small scale on which men and women are known to be at their most comfortable and creative.

It involves what Dame Barbara Ward Jackson felicitously called 'instructions for the care and maintenance of a small planet', and it addresses itself squarely to most of the problems with which contemporary society is afflicted. It sees none of those problems as discrete or local. Few if any of those who appeared as objectors at the Windscale inquiry believe that the THORP project *alone* is potentially calamitous. Instead, they attempted – some explicitly, some less so – to weave their opposition into a broader argument about resource-management and the reconciliation of man's aspirations and abilities with the constraints and vulnerabilities of his natural environment. Their case, simply, is that THORP and all that is predicated on the construction of THORP represents one more in a number of misguided steps.

There is no recognition of this in the Parker report. Parker and his two assessors did what was narrowly expected of them by their 'client' – the Department of the Environment. It concerns itself with the technical, fiscal and specific institutional aspects of the planning application called in for the inquiry. It is a brisk, businesslike, steadfast and unambiguous rejection of the opponents' case – and has met with broad initial acceptance from the Government, the civil service, a large proportion of the general public, the press and – naturally – the industry.

The report's first substantive section is that dealing with the question of nuclear-weapons proliferation – rightly regarded by many as a key issue, if not *the* issue, on which the argument for the reprocessing of reactor fuel stands or falls. (This is examined from some of the opponents' point of view in Chapter 5.) Parker begins with a pre-emption, essentially, of the argument that Britain should not separate and export bomb-making grades of uranium and plutonium because the USA has now taken a stand against international commerce in these materials. He quotes one witness – Professor Albert Wohlstetter – and the energy analyst Amory Lovins to show that the USA has exported considerable quantities of strategic fissionable material, up to 1976, to non-weapon states, and observes that these exports were made on the understanding that peaceful purposes only would be served and that relevant safeguards would be followed.

He concludes that 'These undertakings, so far as is known to me, have been honoured.'

By implication, the report is saying that the American position on reprocessing is unreasonable, since that country once did what Britain is now asked not to do; and that, in any case, no real risk can be seen to have been taken. Since the word proliferation means what it says, the conclusion is, at best, an ingenuous one. The USA, it could be argued, belatedly but wisely acted in the spirit of the nuclear Non-Proliferation Treaty (NPT) and it is up to the UK to follow that initiative. The report acknowledges that the NPT is in need of improvement, but does not concede that any nation has not the right to (a) develop and use reprocessing for the production of plutonium; (b) develop and use the FBR; (c) have access to the technology and equipment for creating reprocessing facilities; or (d) have access to reprocessing facilities in another's territory and to the plutonium produced by those facilities.

The NPT, says the report, is a straightforward bargain: nuclear-weapon states will afford every assistance to non-weapon states in exchange for the necessary undertakings on proliferation. Here, it introduces or grafts on the objectors' argument about the possible economic loss represented by the construction and operation of a THORP: 'Such expense or loss is a natural price for securing the undertaking from non-nuclear-weapon states not to become such states.' This response is a strange one to points raised in cross-examination – not always explicit points – rather than in submitted evidence. The objectors were not saying 'All right – the NPT obliges us to sell plutonium overseas, but we shouldn't because, apart from anything else, it will be financially unrewarding.' Instead, they were saying that the NPT is a device for stopping the spread of nuclear weapons, *not* one for helping the commercial and technological intercourse of the civil nuclear-power industries. Challenged on the latter point, they merely restated their contention that net fiscal national benefits from such intercourse are unproven. The distinction is an important one.

After a brief rehearsal of parts of the NPT, the report goes on to deal with the question of American policy. It quotes – and gives prominence to – some of the remarks made by President Carter at the now-celebrated press conference on 7 April 1977, at which he spelled out his policy for a moratorium on reprocessing and a deferment of the introduction of FBRs. At the conference, Carter said: 'We are not trying to impose our will on those countries like Japan, France, Britain and Germany which already have reprocessing plants in operation.' But Carter went on to declare that the new policy was intended to 'set a standard' for the steady and positive curtailment of reprocessing and expressed the hope that such countries 'will join with us in eliminating in the future additional countries which might have had this capability involve'. Unfortunate phrasing, perhaps – but the meaning is clear.

And yet the report concludes that the building of THORP 'would not be counter to US policy so long as no plutonium produced by it was exported . . . If the use of THORP were not so limited and plutonium were supplied to non-nuclear-weapons states it would not be so supplied until, at the earliest, ten years from now, for THORP would not be operative until then.' Objectors do not think much of this argument; neither do American politicians and commentators. The *New York Times*, in a leading article in mid-March 1978, noted that the Carter administration could, if it chose to, severely limit Britain's aspirations to be 'reprocessor to the world' and warned that the UK could not count on securing US permission to service nuclear fuel (such as that burned in Japanese reactors) which was originally supplied by America. Inferring that the US Government might be biding its time while other diplomatic issues were resolved – such as those concerning Rhodesia presumably – the *Times* nevertheless counselled Carter to 'give a stronger signal of American distress and reaffirm his commitment to block the further spread of bomb-proliferating technology'.

Later in the same month, five US legislators – two Senators and three Representatives headed by Senator John Glenn, chairman of a Senate sub-committee on nuclear proliferation – wrote to Carter, making a direct plea for a renewed policy statement. American opinion, they told him, had been badly misrepresented at the Windscale inquiry (and, by strong implication, in the

Parker report). Concessions intended by Carter to minimize short-term commercial and diplomatic disruption had been characterized as a long-term policy of conciliation. 'The likely consequences of such distortion are grave. An extension of existing European reprocessing plants could shatter our efforts to forge a consensus.'

To other opponents, it seemed at least imprudent to dismiss the proliferation case before the members of the INFCE programme had met to consider the full implications of a go-ahead for THORP. But Parker had even pre-empted that: 'Even if THORP did become redundant [as a result of INFCE], and I do not consider that it would, this would merely mean that some expenditure had been wasted.'

The question of terrorism and civil liberties rates only five pages in the Parker report. One justification for this seems to be that the scope of evidence was, by statutory necessity, limited. So it was: but there is far more to be said on the subject – especially in a document of this sort – than can be compressed into such a space. The points raised in Chapter 5 are distilled from *some* of the evidence submitted to the Windscale inquiry; that evidence itself was largely synthesized from a far greater volume of research and analysis into subversion, blackmail, hijackings, counter-intelligence and surveillance in the nuclear business. But the submission made by Dr Tom Cochran on behalf of Friends of the Earth is only briefly cited in the report – and in such a way as to suggest that Cochran put forward technical answers to the problem of terrorist diversion of weapon-making nuclear material.

At the time of writing – the end of March 1978 – Cochran had sent a formal protest to the Secretary of State for the Environment, Peter Shore, expressing 'shock and dismay at the way the judge misrepresented my testimony to support his own findings'. FoE went so far as to say that, if the Parker report had been a judgement in a court of law, they would have grounds for an immediate appeal. Cochran, they pointed out, had specifically listed the security *problems* to be anticipated from an expansion of reprocessing – not the solutions. The fourteen-line excerpt from his cross-examination, reproduced in the report, completely distorted his position as a leading and informed critic in the USA of reprocessing and the FBR.

Parker accepts that the manufacture of a crude nuclear explosive is feasible, but he goes on to say, baldly, that 'although plutonium has been produced and moved, both intra- and internationally for over twenty-five years, there has not been any terrorist abstraction or threat so far as is known'. He also concludes that there is no evidence that, at present, the safeguarding of plutonium has constituted any 'undue interference with civil liberties'. He says that it was not seriously suggested (by the objectors) that THORP alone would involve any significant interference with civil liberties 'and I do not consider that it would'. He was not asked to. He was invited to consider the proposition that THORP, among other proposed nuclear-development projects, would come at a time when social stability, worldwide, is more likely to have deteriorated than to have improved, and that acts of desperation both by governments and their more militant opponents will have increased in frequency and severity. The inquiry needed no more than a daily newspaper to consider the issue adequately.

The report introduces, or repeats, the 'energy or extinction' argument as it touches upon the matter of civil liberties in a society committed to the use of plutonium-fuelled power stations. It says: 'It might be that [in fifteen years' time] ... short of a large and immediate commitment to FBRs, with whatever erosion of civil liberties might go with it, the country would have to accept a severe reduction in living standards or a greatly increased pollution of the environment from coal-fired stations ... what would be acceptable in the field of erosion of civil liberties must depend on what the alternatives are, or are estimated to be ...'

It is a legalistic nicety. Of course the benefits and liabilities of withdrawing certain rights, freedoms and privileges are subject to debate and have been for five thousand years. The objectors' argument is that we need not pursue technological options that make that debate significantly more difficult to conduct. In the end, Parker is himself pessimistic, hopeless even, on the question of surveillance: '... I can see no solution at all.

If the sort of activities under consideration [i.e. subversion] are to be checked, innocent people are certain to be subjected to surveillance, if only to find out whether they are innocent or not. Equally certainly, friends and relatives will be subjected to distasteful and embarrassing inquiries. The most that one can do, as it seems to me, is to require that the Government should ensure that the interference with our liberties goes no further than our protection demands and that there should be some Minister answerable to Parliament if interference goes further than this.'

This passage, again, seems fraught with interpretive pitfalls – and who is that Minister to whom the report refers but the existing holder of the Home Secretaryship? Objectors can show – and did so during the inquiry – that security measures already do go 'further than our protection demands'. They have to, since such measures must be anticipatory as well as reactive if they are to have effect. Lastly the report takes up the suggestion that those who say they seek to protect civil liberties may have the aim not of preserving such liberties but of increasing the opportunity to further seditious ends. The report rejects the suggestion, though it notes that some witnesses were in favour of a reduced standard of living (a view that might be construed by some as intended to undermine national strength) as an alternative to reliance on nuclear power. There was, it declares, 'no evidence before me which went even a small way towards establishing that the country at large would be prepared to accept such an alternative'.

Here, once more, there is an encapsulation of argument to the point of confusion. That the public might or might not be prepared to accept conservationist policies for the management of energy and other resources has not been tested. And to present them as straightforward alternatives to a nuclear-power programme is to make a risky philosophical short-circuit. An 'acceptable' standard of living can be defined in many ways, but it does not have to embrace the sequestration and grossly inefficient use of raw materials any more than the embarkation on a journey towards fission-produced electricity as a major source of energy. As several observers have pointed out, any list of 'alternatives' drawn up forty years ago would not have included nuclear power. The energy alternatives now being canvassed cover a wide spread not only of technologies but of their social and environmental implications – something which goes almost wholly neglected in Parker's narrative about the need for reprocessing.

Here, the report begins by examining the relationship between a THORP and any future fast-breeder programme. An initial commercial fast breeder (CFR1) could be fuelled from stocks of plutonium separated in the Windscale Magnox reprocessing plant – stocks which could support such a programme up to and including an eighth FBR, possibly around the year 2001. From this it concludes, justifiably, that oxide reprocessing is not essential to keep open the option – over that time scale – of an FBR programme. A THORP would be needed at *some* stage. When that would be depends (a) on whether FBRs were to be built, as suggested, to come on line from 1990 onwards; and (b) how many AGRs might by then be in operation and providing the source of plutonium via THORP. Nowhere is the obvious stated: that, if FBRs were *not* to be built, there would be no need of THORP on the grounds of plutonium production.

The need is reviewed next in the light of argument about what to do with present and planned arisings of thermal-oxide spent fuel. Can spent unreprocessed fuel (SURF) be satisfactorily stored together with its content of plutonium, uranium and waste actinides? The report concedes that no 'final solution' to the problem has been found or advanced, but it accepts the assurances from BNFL that the feasibility of vitrification has been established, and accepts also that the risks from storing separated actinides are lower than those that would present themselves from storing SURF. It says that there will be 'substantially more plutonium against which to protect future generations by not reprocessing than by reprocessing'. But if separated waste can be safely stored, there appear to be no very strong reasons why SURF cannot be so stored, thus removing the risks of estracting, warehousing, transporting and handling plutonium. Going back to the previous paragraph, one might ask what is to be done with extracted plutonium if FBRs are not constructed? The question touches both issues.

Turning to the question of energy conservation and resource independence, the report begins with a tart reminder that 'neither the US nor Canada are presently troubled by the latter problem in respect of uranium supplies. Both have uranium supplies sufficient not only to supply their own needs for many years but also to allow them to use the threat of withholding supplies as an instrument of policy, the effectiveness of which is beyond doubt ... if we are going to depend, for a substantial part of electricity supplies, upon nuclear power, it is in the public interest that we should, unless the price of doing so is too great, minimize reliance on imported fuel.' If uranium and plutonium, recovered in THORP, were to be used in AGRs, then Britain's reactor fuels could be used with 15 per cent greater efficiency than with a single-cycle fuelling; in FBRs, the efficiency of the fuel would rise sixty-fold. To dispose of recoverable fuel, the report says, 'would therefore be an act of folly'.

It accepts that energy forecasting is an uncertain business but argues that, because of this, the wisest course is to adopt a strategy relatively unaffected by the variables so that the energy needed 'to support an acceptable society' can be provided. Is nuclear power likely to be such a strategy? On what basis can the likelihood be determined? At the moment, as noted elsewhere, the financial, technical and intellectual commitment to nuclear power as a major source of electricity is vastly greater than it is to the alternatives. The official books may not be cooked, exactly, but they are prepared in a shadow: if nuclear strategies are *not* the right ones, then we have been wasting billions in money and manpower. That is not easily accepted in places like the Department of Energy, from whose forecasts Parker concludes that 'it would be imprudent not to continue to develop nuclear technology and keep the nuclear industry in a condition to meet a sudden expansion in nuclear power should it be required, be that expansion in thermal reactors or in FBRs'.

The report says that it is 'recognized' by the Government (i.e. the British Government) that conservation policies and alternative-energy projects should be pursued, but suggests that a significant diversion of research and development funds from nuclear power might be 'an act of bad management for which this

and future generations might justly blame the Government . . .'
It describes claims for a large increase in coal production (made
at the inquiry by a miners' leader, Arthur Scargill) as 'fanciful'
and goes on to reject, in one short paragraph, *all* the alternative-
energy sources presented by witnesses at the inquiry. 'The
witnesses who spoke of them were, on the whole, moderate, but
because the contributions of which they spoke are of necessity
uncertain, it does not seem to me that any detailed consideration
would be of assistance for present purposes.'

The report includes a longer mention of the polluting effects
of coal-fired power stations, mentioning the emissions of sulphur
dioxide, carbon dioxide, nitrous oxides, carcinogenic hydrocarbons, heavy metals and mutagens. No one, as far as I can recall,
submitted evidence to the inquiry suggesting that all was environmentally well with the electricity-generating world we have
at present: Parker turned it into part of an opposing scenario.
He appears to believe that funds would not be made available
for a particular technology if that technology did not show real
promise, and concludes with the mighty assertion: 'Save where
resources are limitless the final development of a large project
will always lead to delays in others, and, if this were not acceptable, no large project would ever reach fruition.' Again and again,
the objectors' central message would appear to have been lost:
Concorde, the centralized production of steel, the 'rationalizing'
of the railways, the advent of one-trip bottles – they all came to
'fruition'. And they have all been socially, economically or environmentally deleterious.

The conclusions at the end of this section of the Parker report
are perhaps the ones on which all others hinge. They are summarized as follows:

*That thermal-oxide-fuel reprocessing is not necessary for the
purposes of preserving an option to build CFR1 or for the purposes of enabling an FBR programme to be launched. Plutonium
for either can be recovered from Magnox reprocessing.
*That an FBR programme developed beyond eight reactors
would require a THORP or the slower introduction of FBRs.
*That additional plutonium from a THORP would not be re-

quired until after 1987; a start on THORP could, thus, be delayed – possibly for five or even ten or more years.

* That, if delay were countenanced but reprocessing ultimately required, the work would have to be proceeded with at an accelerated rate, which would involve greater technical risk.

* That, if deferment were accepted, long-term storage of SURF would be required and would necessitate an urgent programme of R&D.

* That storage of SURF involves greater environmental risk and the throwing-away of energy sources.

* That there is little to be gained from such R&D because 'it is undesirable to dispose of spent fuel without reprocessing'.

(The logic of these last two declarations is unfathomable.)

* That the nuclear industry should be kept in a state of readiness for further expansion.

* That this need might be precipitated by recognition of the fact that coal-fired power stations are a probable source of greater environmental harm.

* That there is a world need for reliable reprocessing capacity.

Taken together, these points lead Parker to conclude that 'there should be no delay in building the plant'.

The report makes one important concession to the objectors: it sees as justified their criticisms that BNFL had not done its financial homework in presenting the case for THORP. But Parker rejects the criticism as being invalid. Apart from finding, for instance, Dr Colin Sweet's detailed counter-analysis 'unconvincing', he does not accept submissions on the fiscal worthiness of THORP because, firstly, he believes it to be procedurally inadmissible; and, secondly, he contends that 'financial disadvantage might be an acceptable price for some other advantage, for example resource independence, reduction of plutonium stocks, or anti-proliferation effect'. That BNFL or the Government agencies had argued nothing of the sort is not revealed in his treatment of the point.

On routine environmental discharges from the proposed plant – those which should fall within the limits laid down by national and international radiological agencies – the report is steadfast. It does not accept that, if there are serious differences of opinion among experts in the field of radiological protection, such differences are reasonable grounds for refusing or deferring the granting of permission for a plant such as THORP. There were, as the report observes, three ways in which it could be shown that the environmental-protection system might not be reliable: if THORP's operation would involve the release of radiation at intolerably high levels; if the system itself was defective; or if the 'competence, capability, or integrity' of bodies making up the system was in doubt.

The report reviews those bodies and their interrelationship, noting that the principal one concerned is the International Commission on Radiological Protection (ICRP), which recommends – but has no powers to fix – radiation limits, adherence to which it considers will sufficiently protect human beings from harm. It is independent of any government, and its members are chosen on the basis of their scientific reputation. The ICRP does not recommend limits designed to protect the general environment: '*It considers that if man is sufficiently protected, so also will be vegetation, birds, beasts etc.*' (emphasis added).

The other major bodies involved are (1) The UN Scientific Committee on the Effects of Atomic Radiation (UNSCEAR); (2) the International Atomic Energy Agency (IAEA); (3) the World Health Organization (WHO); (4) the Food and Agriculture Organization (FAO); (5) the Nuclear Energy Agency (NEA) of the Organization for Economic Cooperation and Development (OECD); and (6) EURATOM, which does set binding standards based on ICRP limits. As far as the UK is concerned, the UN agencies (1–4) and the NEA have no more than advisory status. Under the recommendations made in the Royal Commission Sixth (Flowers) Report on nuclear power and the environment, responsibility for nuclear-waste management in Britain is being transferred from the Department of Energy to the Secretary of State for the Environment and the Secretaries of State for Scotland and Wales. Up to now, it has been divided between a number of Departments, including Energy. A Nuclear Waste Management Advisory Committee is being established for Government consultation.

The position will, however, still be a complex one, with many semi-autonomous agencies involved in research and monitoring. Responsibilities are spread among the National Radiological Protection Board, the Ministry of Agriculture, Fisheries and Food, the Fisheries Radiobiological Laboratory, HM Alkali and Clean Air Inspectorate, the Medical Research Council and the UKAEA. In addition, work is conducted in universities in Britain and around the world. Not surprisingly, the volume of published material on the subject – in the report's own words – 'can only be described as enormous'. The crucial question is whether the quality of that material matches the quantity. There are those who think that it does not.

What they allege is not a conspiracy of wilful negligence so much as a systemic inertia, which was referred to by a number of witnesses at the Windscale inquiry – notably by Dr Brian Wynne and Dr Edward Radford. Radford, professor of environmental epidemiology at the University of Pittsburgh, was concerned to show that the cancer risks to members of the public exposed to certain radionuclides were greatly underestimated; for americium and plutonium he considered the maximum permissible concentration should be reduced by a factor of 200. He made other specific criticisms in his lengthy submission, some of which went by default through the lack of documentary evidence to support his testimony. Though not an opponent, *per se*, of nuclear power and able to praise much of ICRP's work, he strongly suggested that the Commission suffered from serious institutional faults:

'It is evident to me that the small group of ICRP scientists who, over the years, have made the judgemental decisions on which the occupational and non-occupational standards have been based have only slowly realized the complex problems that must be faced in defining standards and their criteria of acceptability. This is amply illustrated by the style and content of ICRP 26 (the Commission's 1977 recommendations) . . . in the social milieu represented by membership on the ICRP it is inevitable that proposed new members will be judged as acceptable on the basis of the weight of opinion prevailing among voting members.' The ICRP, said Radford, was a self-perpetuating group responsible to no political entity, but their scientific function had been gradually blurred into that of making policy.

The report recognizes the force of some of this argument (although it does not address itself directly to Radford and Wynne). It was, says the report, 'repeatedly stressed to me by objector after objector that, in the nuclear field, developments in technology, in radiobiological knowledge, in public attitudes, and in policy change with great rapidity. It did not need stressing. The situation is very obvious. Within the next few years much may happen. Present limits may be reduced so that what appears safe now will be accepted as unsafe. Or they may be relaxed and what appears as unsafe now may be accepted as safe . . . In such a situation, no final answer to the question is possible . . .' Apart from the impatient tone, this is an unexceptionable synthesis of the position.

But Parker, later in the same section, *does* answer the question, and does so to the satisfaction of the industry and the agencies concerned. He accepts, he says, that a largely self-elected body may tend to perpetuate its own thinking, 'but it need not do so, and the fact that standards have remained unchanged is no evidence that ICRP has done so. It is neutral. Standards would remain unchanged if they were correct when last recommended or if they were then too strict but the Commission considered that, since it had proved possible to comply with them, it was unnecessary and undesirable to relax them.' The ICRP, he felt, was a sitting duck, open to accusations from all sides. On the whole, he thought, the checks were 'adequate'.

Parker looks at several of the submissions made specifically about the routine discharges from Windscale and those that would be expected from THORP. He recommends the setting of specific limits for each significant radionuclide, with the onus on BNFL to show that such limits can be met, and he also recommends that the company should pursue R&D on methods of containing krypton$_{85}$. Apart from these, his conclusions are unambiguous: 'I am encouraged to believe that there has not been, is not, and is not likely to be any real cause for alarm.' As we saw earlier, he singles out one witness – Dr Alice Stewart – for special and prolonged attention. As this book went to press, her studies on occupational cancer risks were being evaluated yet again, and new evidence was emerging, from an analysis of American workers exposed to radiation in the Portsmouth naval dockyard, of greater occupational risks than have been accepted hitherto.

The report rejects any suggestion that research or its results have been kept within the confines of those institutions whose interest it is to promote the growth of the nuclear industry or to control its operations. It is even stronger on the question of integrity: 'I have no doubt as to the integrity of those concerned in all of them [the regulatory agencies in Britain] and I regard the attacks made upon them as being without foundation. Such attacks did nothing to further the cases of those who made them and at times reached a level of absurdity which was positively harmful to such cases.'

On the risk of accidents at Windscale or those involving radioactive material in transit, the report is brief and reassuring. There is, as Parker makes clear, little or no danger of a nuclear explosion in a reprocessing plant: none of the objectors sought to show that there was. He records that 'there were paraded before me a range of possibilities, including even the shelling of the site in a civil war and aircraft crashing on to the site ...' Many would leave out the words 'paraded' and 'even', for what some objectors were attempting show, not unreasonably, was that such events are possible and that, if they occurred, radioactivity could be released. An accident need not necessarily be a mistake: terrorist bombs have been exploded on Spanish and South American nuclear sites; a plane hijacker has threatened to dive the craft on to a US reactor.

But the report confines itself to the possibility or probability of orthodox accidents – risks of the kind programmed into the computer routine run by the Safety and Reliability Directorate (SRD) of the UKAEA. Their so-called TIRION programme simulates a discharge to the atmosphere following a failure of an HAW's cooling system. It assumes a release of one ten-thousandth part of the radioactive content of the tank – a release of some 10,000 curies to the air. This, according to the

TIRION run, might lead to the contraction of cancer by ten people, the necessity to evacuate for a few days people living within a mile of the site, and the ban for a few weeks on consumption of foodstuffs produced inside a ten-mile radius.

One body of objectors – the Oxford Political Ecology Research Group (PERG) – asked to be supplied with the source data for TIRION: the Group does not believe it satisfactorily simulates a worst-case accident on the assumptions given.

Parker, as mentioned elsewhere, does not accept the argument that 'if it can happen, it will happen' and sees added cause for assurance in the Flowers recommendations for a review of the workings of the Nuclear Installations Inspectorate (NII). Together with 'constant self-reminders' in the company as to the existence of risk, it is his view that THORP can be built and operated to tolerable levels of safety, both for its working staff and for the general public. As for transport – of fuel, plutonium or waste – he has no reservations whatever. He does, though, recommend that the local emergency plans be made more explicit (which they were in April 1978) and that the Windscale liaison committee should be made a fully democratic body whose proceedings are made public.

The next two sections of the report – 12 and 13 – are of conspicuous brevity. In the case of the first of these, a compact summary was to be expected: one major witness, Dr Barry Shorthouse of the Open University, had challenged BNFL's proposals on the grounds that they were unlikely to be able to proceed satisfactorily from a 1 : 5,000 scale pilot plant to the full-size THORP of 1,200 tonnes annual capacity. Divorcing the issue from the question of general industrial accidents and failures in plant design, BNFL and their supporting witnesses were able to refute this submission as a substantive objection to the plant.

Section 13 deals with public hostility, and deals with it in two pages – more than a third of which space is concerned with questioning the technical honesty of an exhibit brought before the hearings – a film by the former nuclear engineer Dr Charles Wakstein, well known as a combative scold of the industry for

137

what he alleges are its inabilities to predict and avoid accidents. The film, 'Caging the Dragon', contains two sequences which purport to depict consequences of such accidents: in one case a fire; in the other a radiologically mutilated body. Both, it transpired, were from other contexts. They were construed by Parker as likely seriously to mislead the average viewer of the film and as 'particularly undesirable in a film made and shown by a professional engineer'. The strictures are no doubt in order, but the weight attached to them in this section seems, to say the least, undue. Most observers would have expected far more attention to be paid to the submissions made *specifically* about public hostility to and acceptability of nuclear power and re-processing.

The report does not get down to examining properly (a) what the reasons are for such hostility or (b) what account should be taken of that hostility, whatever its merits or motives. Parker observes, fairly, that there is no doubt that hostility and anxiety exist, but says: 'What is the strength of the hostility or anxiety and in what proportion of the public it exists are, however, matters which I was unable to assess with any accuracy.' And yet he feels able to say, at the conclusion of this section, that '. . . no weight at all can be given to the results of a number of opinion polls or petitions to which I was referred'. It is a somewhat astonishing statement, to which there are two more to be added.

Much of the opposition, says the report, appeared to be based on sincerely held moral grounds: '. . . amongst those who advanced opposition on such grounds there was a tendency to suggest that supporters were acting in an immoral way' – and it goes on to speak of the possible benefits of nuclear development set against the possible liabilities of non-nuclear developments. But that, if there was a moral argument, is not what it was about. Very few of the 146 witnesses who appeared before Justice Parker as objectors made non-scientific submissions. The significant 'moral' statements were almost all made within rigorously informed and prepared proofs of evidence, such as those from Brian Wynne, Peter Taylor, Dr John Davoll, Albert Wohlstetter, Joseph Rotblat and several others. In none of these was there a suggestion that the proponents were acting immorally; rather that they were acting unwisely and were victims of their own social and technological momentum.

The remark shows a signal lack of recognition of what has been happening in the scientific community and in industry over recent years. Unquestioned acceptance of scientific and technological orthodoxy, whether in the use of experimental animals, the research commitment to military-hardware development, the deployment of chemical-intensive agricultural techniques – or the trend towards a nuclear-electric energy economy – is with us no longer. Parker's two assessors, Sir Edward Pochin and Professor Sir Frederick Warner, are surely aware of the existence of the British Society for Social Responsibility in Science, the Council for Science and Society or, in the USA, the Union of Concerned Scientists. It does not show in the report they helped to assemble.

Later in this part of the report, Parker makes a curious inference, which, to paraphrase, is this: if Cumbria County Council – the local authority within whose jurisdiction the THORP application initially came – could eventually be persuaded that the application should be upheld, then no greater weight should be attached to the views of those who, at the beginning of the hearings, were opposed. That is: they should have changed their minds in favour of the proposed plant given that they were now in possession of the same information as were the County Council. This is not an assessment of public acceptability so much as the movement of a very rash chess player.

The conventional planning issues are dealt with in more or less the proportions they consumed of the inquiry's time. The report concludes, to no one's great surprise, that, if a reprocessing plant is built at all, it should be built at Windscale, where the expertise, experience, special skills and basic plant facilities already exist. Windscale, it says, 'is a suitable and proper site, indeed *the* proper site'. There is no comfort for those who – like the Town and Country Planning Association – argued that a decision could not be reached without applying an environmental impact analysis (EIA) or assessment, as is common in the USA. The report states clearly that, in this case, there was no legal requirement for an EIA and that the absence of one did not constitute grounds for refusing planning permission.

The general conclusion on THORP is that, in planning terms, there is no substantial objection and that it would be likely to bring some positive benefits into the area; it makes the rather obvious point that, had the proposal been for a factory making, say, machine tools, it would not have encountered overall opposition. On the straightforward planning issues – such as landscaping, provision of essential infrastructural services and conformity with statutory regulations, it lays down detailed conditions for the applicants to satisfy – development and construction timetable, visual-amenity requirements, the meeting of appropriate NII and Health and Safety Executive standards and the reporting procedure for accidents. Subject to these, it is recommended that permission should be granted.

The principal recommendations are these:

(1) Consideration should be given to charging some independent person or body with the task of (a) vetting security precautions both at Windscale and during the transit of plutonium from Windscale, and (b) reviewing the adequacy of such precautions from time to time.

(2) BNFL should devote effort to the development of plant for the safe removal and retention of $krypton_{85}$ and, if development proves successful, should incorporate it in the proposed plant.

(3) More permanent arrangements for whole-body monitoring of local people should be instituted. Subject to certain principles, the details should be agreed by those principally concerned. They would not be appropriate to planning conditions.

(4) The authorizing departments should, however, consider whether provision of such facilities should be made a condition of authorizations to discharge.

(5) Consideration should be given to the inclusion of some wholly independent person or body with environmental interests in the system for advising central government on the fixing of radiological protection standards. That person or body should probably be changed from time to time.

(6) A single Inspectorate, as recommended by the Royal Com-

mission, should be responsible for determining and controlling all radioactive discharges.

(7) There should be specific discharge limits for each significant radionuclide. The onus should be placed clearly on the operator to show that before the limits are fixed a discharge cannot practicably be avoided.

(8) The provisions of the Radioactive Substances Act 1960 relating to the powers to hold inquiries into proposed authorizations to discharge should be re-examined.

(9) The relevant authorities should carry out more monitoring of atmospheric discharges.

(10) FRL should publish their annual reports more rapidly in future. There should, as recommended by the Royal Commission, be one comprehensive annual survey published of all discharges and, at intervals, reports by NRPB on radiation exposure.

(11) BNFL should do more, in future, to ensure that safety precautions and operating procedures at Windscale are sufficient for all eventualities, are strictly observed and are continually rehearsed.

(12) The current review of NII should examine whether they are sufficiently equipped with scientific expertise to check the designs for the proposed plant.

(13) It is essential that those who would be required to take action under the Windscale emergency plan are fully aware of the responsibilities the plan places on them.

(14) The local liaison committee should be reorganized and its functions redefined.

(15) Fuel flasks should, as far as possible, continue to be delivered to Windscale by rail, but this is not a matter appropriate to planning conditions.

(16) Outline planning permission for THORP should be granted without delay, subject to conditions.

10 A Conservationist's Case

Most of the submissions to the Windscale inquiry, not unexpectedly, dealt with specific aspects of the planning application entered by BNFL for the proposed reprocessing plant. With no precedent for a public inquiry in which global questions could be raised, it was natural that the objectors should feel their efforts to be best directed at challenging individual claims and assumptions about THORP. One who brought a larger picture to the exhibition was Dr John Davoll, director of the Conservation Society and formerly an executive scientist with the pharmaceutical company Parke Davis. For the past seven years, since he resigned that position, he has made a particular study of the broader implications of social, political and economic policies as they affect man's environment and the natural ecology. Diffident and not given to making gladiatorial appearances in the conflict between industrialists and conservationists, he is nonetheless one of the most respected and authoritative observers of the environmental scene, and his opinion is frequently sought. With his self-effacing permission, I reproduce the evidence he offered to the inquiry.

'I propose to deal with one crucial statement from the evidence submitted by Conningsby Allday (managing director of BNFL) on the need for the proposed reprocessing plant at Windscale. Paragraph 28 of his evidence states that:

It is clear, therefore, that, for the major uranium importing countries at least, the need for reprocessing and the consequent ability to minimize uranium imports through the recycle of uranium and plutonium remains an essential part of their need for the growing development of nuclear sources.

'The "need" for the growing development of nuclear sources

does not exist, in the sense that the problems this growing development promises to solve or alleviate can either be dealt with by non-nuclear alternatives, or are not amenable to technical solutions alone, nuclear or otherwise. These problems include, principally,

(1) the provision of energy to maintain adequate levels of material production, heating, lighting, etc. as oil and gas are depleted; (2) the maintenance of material economic growth as a way of alleviating social stresses arising from inequities and differentials, and from a mismatch between aspirations and possibilities; (3) maintenance of material economic growth as a way of resolving international tensions between rich and poor countries; and (4) the management of the economy so as to provide stable operations at reasonably low levels of unemployment.

'What are "adequate" energy supplies? It is often argued (for example by Sir John Hill, chairman of the UKAEA) that there is a linear relationship between per capita consumption of energy and per capita Gross National Product (GNP). If it is further assumed that increase in GNP is an appropriate measure of a nation's progress, it follows that an increasing supply of energy will be needed, either indefinitely or until levelling out occurs at some value well above present per capita consumption in the United States (which still strives to increase its GNP). Finally, it is argued that this aim can be achieved by massive development of nuclear power and not otherwise.

'The evidence for "a linear relationship between GNP and energy consumption" typically takes the form of graphs – such as that used by Earl Cook in "The Flow of Energy in an Industrial Society". However, since this refers only to commercial energy consumption it gives a misleading picture for the poorer countries, in which a large proportion of the energy actually consumed is derived by burning wood or dung. When one considers only developed countries such as the UK, the points on the graph are widely scattered, with some countries needing twice as much energy as others to achieve a given level of GNP. Even when allowances have been made for differing conditions between countries, the performance of the UK is abysmal, and it would appear more sensible to use our available skills in improving it rather than in providing ever more energy to pour into this leaky vessel.

'Even among those industrial countries that show a better performance than the UK in the efficient use of energy, and in the matching of the quality of energy to end-use (for example by not using electrical-resistance heating for space heating), there is every reason to believe that further improvement is possible. This is only to be expected of economies developed at a time when energy was cheap enough for care in its use to seem unnecessary.

'Turning next to the significance of GNP as a guideline, it may be conceded that, if GNP measures (as it does) predominantly the physical movement of materials, and associated transactions, then a positive correlation between GNP and energy consumption is not surprising. What should not be inferred, however, is a close relationship between GNP and welfare. The economist Professor Oskar Morgenstern has pointed out that

Another trouble with a GNP concept is that it measures, or rather expresses as positive (i.e., as adding to GNP) the malfunctions of the economic system or society. To wit: if we are stuck in one of the thousands of traffic jams, if airplanes are stacked and cannot land on schedule, if fires break out or other disasters occur that require repair – up goes the GNP. More gasoline is used, fares go up, overtime has to be paid, and so on. It would be difficult to find in any other science a measure which simultaneously tells opposite stories of the functioning of a complex system with one scalar number! If we merely improve the scheduling of airplanes and stagger the times of automobile traffic and nothing else is changed – down goes GNP. It goes up, on the other hand, if industry pollutes the air and we create other industries to remove the polluting substances.

Similar doubts about the relationship between welfare and the consumption of physical resources have been mentioned by Alexander Silverleaf, director of the Transport and Road Research Laboratory:

It is increasingly argued that the apparently insatiable demand for transport reflects a fundamental imbalance in society: far from being proud of the ability to move people and goods with ease and at will, our desire to do so may be a symptom of a failure to organize society efficiently and equitably. Perhaps a better-run society would reduce the need for movement and would not use resources, including energy, heedlessly by providing ever more means of transport simply because they were expected to be available.

'It appears, then, that at least a substantial portion of the alleged "demand" for more energy stems from an unwillingness to change systems of organization and evaluation that effectively decouple welfare from consumption of energy. It is reasonable to attribute this reluctance to a combination of inertia and opposition by those with vested interests in the existing system; the latter include, of course, the energy industries.

'It follows from the above that there is no guarantee that an expanding nuclear-power industry will provide the expected increase of welfare. Indeed, if it diverts resources of skill and capital from the more cost-effective routes of energy conservation and increased efficiency of use, and if it conceals the fact that some problems can be resolved only by social and institutional change, it will make matters worse.

'Of all the sociopolitical problems, the most intractable are those which have been relieved by economic growth and especially by the promise that such growth will continue indefinitely. It is the belief that no acceptable political solutions exist for these problems that explains much of the compulsive faith that indefinite growth is possible, and the consequent emotional appeal of the plutonium economy as a painless and almost magical solution for our difficulties. Although in some cases material welfare can be increased with constant or declining energy consumption (for example by concentrating on the maintenance and increase in stocks of goods rather than their rate of flow from-resources-to-garbage), it would be pointless to claim that the necessary structural change will be universally welcomed by all interest groups.

'Growth can be seen as a way of relieving internal social stresses. Although it is often claimed that we need energy in order to provide a higher standard of living, it is not so much the absolute level reached as the promise of an ever-increasing level of consumption that takes the edge off social conflict, by making it possible (at least in theory) to make the poor richer without making the rich poorer – something strenuously resisted by the latter (this point has been explicitly made by the late Anthony Crosland).

'Although the political attractiveness of the more-growth approach is not in doubt, the belief that unlimited growth is possible is now waning. At some point in the near future (since exponential growth is so powerful as a multiplier), problems of equity and distribution, both in the present and inter-generationally, will have to be solved in the context of non-growing or declining material production. Moreover, it seems obvious that to defer action until the last possible moment will be risky at the least.

'A supplementary point to this is that, for reasons explained by Fred Hirsch (author of *The Social Limits to Growth*), it is impossible to expand the availability of certain "goods" such as freedom from interference by others, unspoiled natural environments and so on. Indeed, the increasing production of material goods may actually increase certain types of deprivation.

'Again, growth could be seen as a way of relieving inter-nation stresses. It is often claimed that abundant nuclear energy is essential if the poor countries are to develop along the lines that the industrial nations have already followed (Sir John Hill, chairman of the UKAEA, makes this claim). One argument to this effect is based on the now largely discredited conventional wisdom of twenty years ago that continued economic growth among the rich countries would draw the poor along with them to eventual affluence. With the failure of these expectations, there is more recognition that any significant move towards global equality – and a more peaceful world – will necessitate restraint by the already rich. It follows that the provision of ever-growing energy supplies to the rich countries (as envisaged in most projections from the nuclear industry) is irrelevant, or worse, to the over-arching "north–south" confrontation that increasingly dominates the international scene.

'It has even been argued that rapid development of nuclear power will enable rich countries to leave oil as a cheap energy source for the poor countries. It is hard to imagine a more unlikely eventuality, if past experience of the behaviour of the rich countries is any guide.

'Lastly, growth can be regarded as an aid to managing the economy. Under the present system, economies have undoubtedly proved easier to manage during periods of growth. Even under these favourable conditions, however, intractable problems such as inflation and structural unemployment have arisen; indeed, built-in expectations of open-ended growth could well have contributed strongly to inflationary trends. First, it is not obvious that diversion of large amounts of effort into nuclear expansion is the best way of dealing with the problems mentioned above. Indeed, a move towards an "alternative" non-nuclear strategy appears more promising for this purpose. Second, since an accommodation to a non-expanding economy must be made before long, a timely adaptation is advisable. Once again, nuclear expansion could serve as a hindrance; far from being genuinely innovative, it represents an attempt to avoid basic changes in the industrial system.

'As far as the "energy gap" is concerned, it has been claimed that nuclear power must be expanded in order to meet energy demand in the 1980s and 1990s as UK and world oil reserves are depleted. The forecast of an energy gap needing to be filled depends on projections of rising energy demand in the UK with continuing economic growth. In their turn, these imply increasing consumption of raw materials as well as of energy, and appear to ignore the enormous pressure on resources if the rest of the world's economy is growing in the same fashion as is ours. In addition, they take no account of the consequent widening of the gap between rich and poor countries if the growth could be achieved.

'Even if resources of capital and skills were available for a very rapid expansion of nuclear electricity generation, it would not substitute for raw materials, or even other fuels. For example, in the time left before the road-based UK transport system runs into serious trouble from shortage of oil-based fuels, there seems to be no feasible way of providing the expanding mobility assumed in official projections.

'Advocates of non-nuclear features are often challenged to "fill the energy gap" [they were specifically requested by Justice Parker to outline their proposals for alternatives, which they did but which went ignored in the Parker report]. It should be emphasized that the ability of nuclear power to do so is quite unproven, in contrast to the technology for energy conservation, efficient use, and end-use matching for energy quality. The application of this technology, combined with some gradual changes (in, for example, transport and the location and scale of industry), should be capable of reducing expected demand to a level at which the use of fossil fuels can provide a "bridge" to the predominant use of renewable resources, such as solar, wind and wave energy. In spite of relatively little official support – in comparison with nuclear power – several renewable sources appear to be technically promising. Once again, it must be remembered that the plutonium-based fast-breeder economy is by no means tried and tested.

'Conclusions:

'(1) The supposed "need" for an expanding nuclear economy is based on a projection of trends established in an atypical period of cheap energy and rapid material growth.

'(2) All reasonable requirements of an advanced culture can be met without recourse to nuclear energy and with static or declining energy consumption if our technological ingenuity is directed towards the achievement of elegance and parsimony in managing demand, rather than towards the blind expansion of supply. In these circumstances, it should be possible to achieve a substantial transition to the use of renewable energy sources in an energy economy that can be safely generalized to developed and undeveloped countries alike.

'(3) Very difficult problems of equity and distribution, both inter- and intra-national, and of economic management, will need to be faced whether or not energy supplies are increased. There is no evidence that an attempt to increase nuclear-energy supplies will relieve these problems for more than a short period if that.

'I conclude, therefore, that the expanding nuclear economy, of which the proposed reprocessing plant is an integral part, represents a gross misdirection of resources, and that this application should be rejected ...

'Because politicians seldom look into the future beyond the next election, they are able to ignore the obvious fact that continuous growth (of population and of GNP) cannot be a permanent feature on a planet which is limited in its space, its irreplaceable resources, and its capacity to absorb pollution and exploitation without ecological collapse.

'Even those who do look a little further ahead are usually content to excuse inaction by accepting the claims of technological optimists that all our problems can be solved by such means as the exploitation of the oceans and the use of nuclear energy. These "solutions" ignore the probable effects on the ecology of the oceans, on the one hand, and the steady accumulation of radioactive wastes on the other, and in any case merely delay briefly our ultimate confrontation with the limitations of the planet.

'The central aim of the Conservation Society is to remove the shackles of outdated ideas from our minds, so that instead of deforming our personalities to try to fit a crowded and mechanized world, we begin to shape a way of life that meets human needs now, and does not foreclose the choices of our descendants in the future. This can be done only if we live within the renewable resources of the earth, and not beyond their limits.'

Nowhere in the Parker report is there a mention of Dr Davoll's submission, and there are only scanty and dismissive references to some of the arguments put by him and others at the inquiry. Cross-examination of his evidence was brief – and an excerpt illustrates the micro-response to what were macro-points: it also gives something of the flavour of the hearings. To those accustomed to following civil and criminal court proceedings, the forensic techniques are familiar. To most people, they are not – one reason why so many eminent witnesses emerged with the feeling that their submission had been misunderstood and their arguments exploited unfairly. Sadly, that is a fact of adversarial legal life.

LORD SILSOE [*QC for BNFL*]: Sir, in my cross-examination I have in mind to refer to Dr Davoll's own exhibits, not all of them. Also, sir, to the proof of Mr [*Dr*] Chapman and to his book *Fuel's Paradise*, which is, I fear, BNFL 253 and, sir, conceivably to the address of President Carter to Congress, 235, sir . . .

Q. Dr Davoll, I would like to ask you this by way of preliminary, if I may, first of all. What do you say should be done with the spent fuel from the AGRs which are in commission and coming into commission in this country?

A. I believe that if the nuclear programme were wound down in a controlled manner, the spent fuel from those plants could be handled, if necessary at the [refurbished head-end plant].

Q. So, point one then, you would anticipate that they would be reprocessed without a new plant?

A. I don't really wish to express a view on that because the choice between reprocessing and not reprocessing requires an expertise that I don't possess.

Q. So your evidence in respect of the arisings from the AGR stations, and I will have to take up with you the question of running them down, but whatever they be, your evidence would not necessarily be to suggest that the spent fuel should not be reprocessed?

A. No, I think if it were decided to reprocess them then I think the facilities I suggest could be used.

Q. Well you are not an expert on this?

A. No.

. . .

Q. I must take the matter a little bit further perhaps on the wider aspects. You don't foresee presumably electricity ceasing to be used as a form of useful power in the United Kingdom?

A. No.

Q. Have you any forecast which you would wish to make on whether the extent of its use in this country could be expected to rise or fall or stay roughly constant?

A. I would rather make a recommendation than a forecast, and I would suggest that if electricity is only used as a source of energy for those applications in which electricity alone will serve, then I think that demand for electricity can be met with co-generation of heat on a perfectly satisfactory scale but with existing plant, with a certain amount of fossil-fuel usage, and gradual imputs from renewable sources.

Q. In such a future, would it be either your forecast or your recommendation that there should be, after the year 2000 or so, no reliance on nuclear-generated electricity, or would you see, quite properly, that there would be some reliance on nuclear-generated electricity?

A. I would prefer to see no reliance on nuclear-generated electricity because I have the fears about a worldwide nuclear economy that were articulated by many other objectors at this Inquiry.

. . .

Q. . . . in meeting political difficulties you would deny society opportunity of assisting meeting them by reliance upon uranium as an energy source save only to a very, very small extent?

A. I think not. I have explained that whatever we do with nuclear power there will be no way of meeting the requirements for hydrocarbon fuels, and if you suggest the electric vehicle I would again quote from [Silverleaf of the TRRL]: 'A recently detailed study of the size, cost, and complexity of such a system of electrical transport has shown that the equivalent cost of electrical fuel, including capital and running cost of the service stations, would be twice to three times that of petrol.' Now I would suggest that nuclear power is not going to have any relevance at all to meeting the crises that will come on the British transport system with the rising price of oil.

. . .

Q. . . . You refer [to Anthony Crosland's views] . . . I am not quite clear whether you are suggesting Mr Crosland's article supports your views or is in conflict with them?

A. No, I merely quoted Anthony Crosland to show the tremendous political pressures on anyone in office to support growth, because Crosland, I thought, was very realistic in his assessment, and I am simply suggesting here that growth, even if you like crude material growth within the present system, has enormous political attractions, and I go on to say so. That is the only point of the quotation.

Q. So as a forecaster of the achievable you would attach weight to the views of Mr Crosland here, would you?

A. As a forecaster of the politically achievable in unconstrained situations I think Mr Crosland is basically right.

Q. In political situations?

A. It may be we shall be put in a situation where we do not have the choice. What I am saying is that if you have the choice of going for growth or voluntarily abandoning it then I think the choice will always be in politically realistic terms to go for it, but I foresee a situation when, in circumstances of which we have no control, we shall be unable to grow in the conventional sense, and I would like us to begin to adapt to that before it comes on us with very unpleasant political consequences.

Q. Absolutely. I follow that, but additionally as I understand it you would hamper that transition by denying the country and indeed as far as you could the world the opportunity of relying upon nuclear power to assist in that transition?

A. No, that is not the case at all, because nuclear power promises rightly or wrongly to supply a very large amount of energy, and energy shortage is not the only constraint, but there are a large number of material constraints as well, because, contrary to some textbooks, common rock plus energy does not equal resources to meet all one's needs, and we shall find we have energy and a very large investment of our capital and skills and finance in a nuclear industry which has no resources to process, and has starved other developments which could have adopted this to a less resource-demanding society in a more painless manner.

Only one other participant wanted to cross-examine Dr Davoll on behalf of the applicants – Dr Kittie Little, a former materials chemist who appeared at the inquiry for Ridgeway Consultants, which she described as 'a dormant consulting company that has not done any consultations up to now'. A constant distraction, even for BNFL, her submissions divided roughly into two parts: that the hazards associated with radiation were largely mythical; and that the opposition to nuclear power springs from a powerful and inspired plot. In her own opening statement, she referred to a US Labor Party paper 'Who Makes Public Opinion?', which links Friends of the Earth and the United Nations to Wall Street financiers, leftist fellow-travellers, and the plan to create 'a deindustrialized slave-labour economy'. BNFL's opponents she persistently characterized as the dupes or architects of a worldwide conspiracy.

Q. The point I am trying to make is that if it is agreed that we should not defer action [i.e., a planned transition to a sustainable industrial system] until the last possible moment, then I would suggest that the only proven alternative is the nuclear ... I do not see how you could disagree with that?

A. You do not think energy conservation is proven?

Q. Energy conservation will not run a steel rolling mill?

A. No, but if you use your fossil fuels to provide process heat and cogeneration of electricity and if you apply conservation so that you need less of one to give you more of the other, you will find that it is much easier working from well-known technology to solve these problems than it is by assuming that breeder reactors will be reliable over the period of use that is expected for them were –

Q. But do you ...

A. Let me finish. Whatever you say, this is not proven.

THE INSPECTOR: Dr Little, you will get nowhere if you try to enter into arguments with the witness. Your position is to ask a question: if you get an answer you do not agree with, then that is an answer you have to put up with. You will advance nowhere by trying to argue with the witness.

. . .

Q. So that [one answer] would be to use less oil for the purpose for which it need not be used?

A. That would not make more oil available in other areas. The use of oil is complex, and petrol and diesel oils are really the by-products of oil for heating. If you do not have an influx of crude oil because the world oil supply has run into difficulties in meeting demand, you will not be able to divert it from other uses into that. Even if you think, and as I take your implication to be, that you could meet some of the demand from nuclear and thus liberate the oil, that does not work ...

. . .

THE INSPECTOR: Dr Davoll, you were asked [about fast-breeder reactors] and on this matter ... I take it from you what say that you would readily agree that however far there is to go with the FBR there is a good deal further to go with all the other alternative

153

sources. Not probably their fault in the least because they have not had the funds provided, but we have got FBRs working and, as far as I know, we have not yet got to a pilot plant in anything else?

A. I would not entirely agree. If the FBR did work exactly as expected and with the breeding rate of fuel and all other problems which have not yet been fully resolved were solved, then what you say is probably correct. The methods of conservation, cogeneration of heat and power, and so on are well tried and moreover are –

Q. I am not talking about conservation. I am talking about alternative energy.

A. Alternative energy sources, wind and wave, undoubtedly yes. Yes, that has gone further.

Q. The only other question I wanted to ask you was this . . . you end up with the sentence 'This can be done only if we live within the renewable resources of the earth and not beyond their limits.' With renewable resources you are thinking, are you, of the sun, in effect, in one or other of its forms – wind, wave, tide?

A. And of course all the produce of agriculture in the natural world. An enormous amount of solar energy goes into this.

Q. When you are thinking of things like wind, wave and tide you are contemplating that man, by the use of his ingenuity and machinery, converts a natural resource into useful energy and that the natural resource which he is converting is a renewable resource?

A. Yes. It does not disturb the general heat balance of the biosphere in that it is not distorting actual flows in the same way that, for example, adding heat from some other source may do.

[This exchange took place on the sixty-first day of the hearings]

Q. I am sorry?

A. If you are using solar energy basically you are not changing the heat balance of the earth. Whereas, if you generate energy by nuclear power you are adding energy in addition to that already falling on the earth from the sun.

Q. Are you not doing that with coal?

A. With coal and oil? Well, strictly speaking you are bringing forward a debt that was incurred in the past. If it is done slowly it should not damage it, but the difficulty with burning coal and oil at the rates we have been burning them is that the build-up of the by-products cannot come to equilibrium by dissolving in the ocean as they would if more time were given. It is really a matter of the rate at which things are done; whether you are doing it much more rapidly than the natural processes that maintain equilibrium can cope with the changes.

Q. And it would be, would it, because of that you would regard uranium as a non-renewable source even if the fast breeder works?

A. Well, it does get used up of course even in the breeder.

Q. It is making as much as it ... no, not uranium. You are using a little uranium because you have got to have your [U_{238}], but you renew your plutonium by the use of man's ingenuity?

A. No, in the end, you use up *all* the uranium and there is neither plutonium nor uranium left; all of it has finished up as fission products.

Q. But why is that not to be done on your particular argument in respect of uranium at a very slow rate but is acceptable to be done with coal or oil at a very fast rate?

A. I am just as worried by the rapid burning of fossil fuels such as coal and oil as I am by the nuclear economy.

Q. I see. Thank you.

11 Public Participation?

Was the Windscale inquiry a model or a mess? As it closed, the Social Science Research Council announced that its Energy Panel had commissioned a survey 'to study and report on the papers and proceedings of the inquiry, with particular reference to the lessons that might be learned for the conduct of future public discussions of energy matters, and indeed of major industrial developments in general'. Professor David Pearce, of the Department of Political Economy at the University of Aberdeen, was awarded the one-year contract and, with assistance from Lynne Edwards and Geoff Beuret (of the Science Policy Research Unit at Sussex University), began work immediately on the Windscale Assessment and Review Project (WARP). An interim report from WARP was due to be delivered to the Energy Panel as this was written and was to go thereafter to the Secretaries of State for Environment and Energy, respectively Peter Shore and Tony Benn.

That report, and the ultimate findings of WARP, will suggest the need for major changes in the procedure adopted for resolving issues such as those raised by Windscale. It is not simply that the environmentalist movement did not get the answers it wanted from the Windscale inquiry: there are many outside the anti-nuclear lobby who have felt for a long time that existing statutory frameworks were inadequate for dealing with contemporary technological questions. Well before the subject of an expansion at Windscale became a controversy, there had been disquiet expressed about the way in which individuals and public-interest groups are forced to fight an unequal battle over planning applications.

David Lock, in a paper written as the Windscale inquiry began, looked at the question of costs. There were, he observed, no national statistics in Britain of the expenses incurred by action groups, residents' associations, amenity societies or other voluntary bodies who contest various applications, 'but enough is known to confirm that there is a very unequal distribution of resources amongst participants at many planning inquiries, which leads to an unequal presentation of the various arguments and, as a result, may distort the judgement of the Inspector or, in the small number of cases that reach him, the Secretary of State.

'When the cost of technical advisers plus their disbursements – postage, travelling expenses, paper, duplicating, maps, photographs, telephone calls etc. – are added together with loss of earnings caused by attendance at prolonged inquiries, a total cost of several thousand pounds can easily be incurred by an ordinary group. Win or lose, there is no provision for them to recoup any of their expenditure, regardless of the contribution they have made at the public inquiry. They are particularly prey to the bleeding of their resources by applicants for planning permission who return to the same site again and again until at last the interest group is both politically and financially exhausted.'

Then there is the question – raised elsewhere – of how genuinely public an inquiry can be. Every account, other than that carried as transcripts of the hearings, is necessarily abridged. Comparatively few people have the time or the motivation to purchase even nominally priced copies of these: they go, for the most part, to those immediately and directly involved and to Government offices, specialized libraries and research institutions. Just as few, perhaps, will be free to attend the inquiry. And as things are, the press and broadcasting organizations will not provide anything approaching full coverage but will select high spots – the beginning and end being the most usual – and orthodoxly newsworthy diversions like outbursts from objectors or the dropping dead of the Inspector.

Thirdly, and possibly most important of all, is the form adopted – the authoritative, hierarchical, party-ordered judgement of evidence heard from a parade of adversaries. Can issues as far-reaching as the proliferation of nuclear weapons or energy policy for the next century be decided in the same way that blame is determined in a civil proceeding for accident damages or a company is examined on the dismissal of an employee? The answer is almost certainly no: and at the moment, there is little likelihood of a truly satisfactory way of arriving at the perfect resolution of such seminal issues. We do not seem to possess the social structure, the discipline, the patience, or – unhappily – the concern to relate and debate them fully. Otherwise there would be no *need* for Windscale-type inquiries: there would be a common and continuing understanding of what were right and what were wrong decisions.

But the procedure could be improved measurably and be applied during the forthcoming inquiry into whether Britain should have a fast-breeder programme. The Council for the Protection of Rural England, in evidence to a House of Lords Select Committee in 1974, looked forward to a CFR inquiry and suggested that an 'examination in public' technique, in which an inspector or panel of inspectors unconnected with any interested Government department and unrepresentative of the industry would be free to conduct the widest possible investigation of the issues and report not to a Minister but to a Parliamentary Select Committee. The CPRE also suggested the use of an Environmental Impact Analysis for certain forms of large-scale development: indeed, the Environment Department itself had asked a study team to look at the relevance of EIAs. CPRE felt (and were vindicated in 1978 by publication of the Leitch report on motorway and road-construction inquiries) that some new form of public consultation and participation was needed if Parliament's role was to be respected. If doctrine continued to prevail on 'policy' issues, then the disruption of statutory procedures would become increasingly common – 'a grim prospect for the democratic process itself'.

In February 1978, WARP invited interested parties to submit their views and suggestions on how future inquiries – especially that for CFR1 – might best be organized. The following areas were tabled for discussion and comment:

1 *Terms of reference:* If an inquiry's terms of reference are an

important determinant of its findings, should there be prior debate over this question? If so, how, between whom, and under what rules?

2 *Format:* Which format would be appropriate to what kind of inquiry? How could function be matched to format? What virtues or problems would there be if objectors were able to nominate one or more commissioners or assessors? Would it be more satisfactory to order the inquiry by topics rather than by parties? Should there be rests/gaps between topics?

3 *Use of lawyers:* If they are to be used, are there other and better ways of using the services of the legal profession?

4 *Timing:* What would be the best way of timetabling inquiries? Could a longer pre-inquiry period be used to: Establish the key issues? Secure co-operation between objectors? Enable all parties to establish their data-base? What are the difficulties in this area? How can they be overcome?

5 *Resource imbalances:* What resources are needed by objectors? Money? Information? Research assistance? Legal aid? An objectors' secretariat? From whom might they be sought? Government? An independent body? Charitable trusts? Could these resources, if they are needed and were made available, be administered or allocated by the objectors? Should the objectors form a temporary umbrella organization for this and any other purpose?

These questions reflect fairly accurately the range of dissatisfactions expressed by objectors during the course of the Windscale inquiry. They had been addressed, formally, by the Town and Country Planning Association, whose director, David Hall, had asked Justice Parker specifically to recommend improvements in the decision-making machinery. The TCPA, of all the bodies represented at the inquiry, was bitterly unhappy about the way in which submissions were timetabled and received: Maurice Ash, the Association's executive chairman, felt that the inquiry did not and could not assess the real alternatives to THORP.

TCPA did not wait for the WARP questions before producing a statement of its own – on 1 February 1978 – 'Energy Policy and Public Inquiries', later submitted to Professor Pearce and his colleagues for consideration in their report. According to this statement, the opportunity for public debate and examination of possible energy strategies is 'non-existent in any sense that would give it credibility'. The Energy Commission (formed in 1977) had published papers in a desultory fashion and with no apparent reference to an overall working programme, minimal publicity or encouragement to comment, and had done so from 'a position of considerable ignorance on some important factors which affect energy demand – such as demographic, social and land-use change'.

The Commission, it added, was too narrowly constituted to produce a balanced view of possible and/or desirable future strategies: it was too dominated by representation from the energy industry, the trade unions and manufacturing industry for it to be able to give anything other than a one-sided view. It suggested that the Commission be scrapped and replaced by a standing body like the Royal Commission on Environmental Pollution and be given broad and flexible terms of reference to advise on national, European and wider international matters concerning energy demand and supply, energy conservation and the adequacy of research into these subjects. It would pay particular attention to alternative energy sources and would take, as its first task, the assessment of a range of feasible energy stategies that might be realistically applied over the next three or four decades. It would have the power to pursue any matter considered relevant and would be asked to produce its first report on energy strategy within two years.

In proposing such a Commission, the TCPA was envisaging that an inquiry into CFR1 would be deferred; indeed, the statement made plain that the first report should be prepared and debated in public and by Parliament before any major energy decision was made, either on nuclear power or, for instance, on the proposed exploitation of coal reserves beneath the Vale of Belvoir. The Royal Commission's energy strategy would form a basis for later inquiries into such development. Thereafter, the Commission would also be directly involved in the inquiries themselves. As for CFR1, the TCPA felt that the most suitable procedure would be that of the seldom-used Planning Inquiry Commission, provided for in the 1971 Town and Country Planning Act.

Such an inquiry would consider not only the consequences and impact of a single fast breeder, but would, ideally, assume that the one would be a precursor to a full-scale FBR programme: 'The whole wedge should be examined, and not just the thin end.' In all this, there would be the need for an EIA – or a series of EIAs for each of a number of possible sites for FBRs – which would examine the total effect on the environment of constructing and operating proposed reactors.

It would clearly be a substantial intellectual and political undertaking – though the precedents in the USA do not look too alarming. Delay is their main characteristic: part of an EIA, for instance, would be the requirement for all information on design, construction and operation to be supplied to the inquiry; all the proponents would need to submit detailed statements in support of their principle. The inquiry itself would not begin until the EIAs had been completed, and even then, it would not enter an adversarial stage. Here, the TCPA suggestions merge somewhat with those submitted to WARP by the Oxford Political Ecology Research Group. PERG's comments begin with the view that a decision on fast reactors should come 'only after long and exhaustive public and parliamentary debate'.

While broadly supporting the TCPA proposals, the PERG responses are generally and specifically much more radical. The group notes that the Flowers Commission was criticized for its ambivalence: 'We feel that this ambivalence shows a proper appreciation of the Commission's role in a democratic society. The Parker report, by comparison, is authoritarian.' PERG raises the matter of public ability to understand the issues involved, and says, bluntly, 'There are a surprising number of people (not always the academically trained) who can grasp complicated material if given access to it. Furthermore, there are numerous channels – including, but not restricted to, the media – through which concepts and opinions can spread, albeit gradually.'

The Group expressed to WARP its concern that all who might be able to speak were not permitted to do so. For instance dissenting scientists and scientists who had resigned their posts from bodies within the nuclear industry were restricted by the Official Secrets and other Acts. PERG would like to see a Freedom of Information Act or reforms based on recent Swedish legislation. Unfortunately, the Government was making it clear, even as PERG was preparing its report, that the British are unlikely to see much if any change in the law relating to secrecy on nuclear and a great many other matters.

A fast-breeder inquiry, in PERG's scheme of things, might take anything up to *four* years and would divide into two main parts with an intervening period of a year or possibly longer. Assuming the existence of a standing energy commission and the empanelling of a Tribunal or a team of inspectors, the format would be as follows:

Part 1
Identification of the issues and areas in need of investigation
– the proceedings would be discursive: the Tribunal or team would meet and receive submissions over a month
– the submissions would describe available evidence, propose evidence to be further assembled, and suggest areas of investigation for the team to undertake
– the team would be empowered to instruct proponents and regulatory agencies to perform investigations (the application lapsing by default) and for objectors to undertake other necessary research
– the proponents would be required to submit, before the inquiry began, data sufficient to establish the full nature of the application.

Intervening period: The team or Tribunal would meet in public, perhaps three times over the period of a year or so, to review the progress of investigations. Further submission could be considered.

Part 2

Testing of the prepared and assembled evidence under oath
– the documentary evidence would be available at least a month before the inquiry opened, although later evidence would be allowed; written proofs of oral evidence would, ideally, be available a fortnight before the scheduled appearance of the witness
– the evidence would be subdivided into issues or topics: this would avoid leapfrogging contentious evidence and would help to reach consensus on non-factual questions
– participants would, nevertheless, be allowed to make opening and closing statements on a party basis
– the team or Tribunal would have the power to suspend Part 2 if a large body of additional evidence was found to be needed
– Part 2 would be expected to proceed 'briskly'

PERG goes on to suggest that such an inquiry could be both generic and site-specific. It would consider all parts of the fuel cycle for an FBR programme, but would also examine in detail – as in TCPA's suggestions – a number of alternative sites. A judge, PERG says, is probably not an appropriate chairman of the team, and they go further to declare that 'There should be no encouragement of lawyers to have any part in the proceedings, as the experience of the Windscale inquiry and the Parker report shows that their training is quite unsuitable.' This, though an understandable reaction, is an impolitic one: lawyers, as a profession, are as good and as bad, and as indifferent, as others – and there were some at Whitehaven who distinguished themselves not merely by their forensic abilities as advocates, but as supporters of and converts to the cause they were hired to prosecute.

The group then suggests a number of proposals for increasing public access to information, with positive involvement of the broadcasting networks and local authorities in disseminating the work of the inquiry. It concludes by supporting the TCPA scheme for a central body which would channel government funds to objectors, and suggesting ways in which the funds might be allocated for research, legal costs (if any) and other expenses. It ends by pleading that eligibility for financial help should not be limited to organized groups: 'Too much institutionalization of the objectors will destroy the public participation we are seeking to create.'

It would be surprising, to say the least, to see even half of these submissions taken seriously by the present or any likely future government in Britain. Inasmuch as it can be gauged, the public mood is not now and has not for a very long time been amenable to letting flowers bloom and thoughts contend (especially when contracts lie in wait for £600 million signatures). The Windscale inquiry was regarded by a majority in both the Government and in Parliament as a horrific concession: to contemplate an exercise which would take perhaps six or seven times as long, consume sizable amounts of public money, and be designed virtually to dash contemporary official plans and objectives, is to dream. Little short of a fundamental shake-up of political leadership, industrial organization and popular public opinion would permit the kind of participation commended in the PERG response. With that sort of shake-up, no doubt, there would be no need either for an FBR programme or an FBR inquiry.

12 Atomic Energy Authority?

Most of the statements made in support of nuclear power in general and spent-fuel reprocessing in particular would meet the recommendations laid down by the Advertising Standards Authority for the tone and content of printed and broadcast commercial messages. That is to say they are legal, decent, honest and truthful. These are wide definitions, and, like advertisements, declarations from the nuclear industry make considerable assumptions of the audience within a similarly broad context. They take it as a granted fact that there is general and *informed* approval of the way in which contemporary industrial society is developing.

They suggest, for instance, that problems like terrorism are to be regarded as discrete: one witness after another, supporting the case for BNFL at the Windscale inquiry, said or implied that terrorism would continue to be a threat whether or not a reprocessing plant was built and plutonium produced. None examined the proposition that terrorism of the scale and nature now practised is the creature of a technocratic world, in which the opportunities for self-determination seem fewer with every new or extended application of scientific and economic logic and in which extreme frustration is a major political component.

What industries like the nuclear one depend upon is an uninformed acquiescence – a negative sanction from the general public for the state and private sectors to control and manage the use of labour and resources. This has long been so, but the dependence has grown as both government and industry have increased their power and influence. The last twenty years are littered with examples of the nonsensical actions and decisions that the modern industrial state *must* perpetrate but which ordinary intelligent people, thinking and working for themselves, would never contemplate.

Here, I am thinking not of highly motivated, voluble and intellectually trained people of the kind who appeared as objectors at the Windscale inquiry, but of the majority – the millions of sensible and responsible men and women whose only power, apart from that exercised by ballot, is also their weakness: their employment. It could be argued long but, I believe, fruitlessly, that organized labour now has very considerable bargaining strength – but bargaining for what? Not for fundamental changes in the nature of employment or the goods and services that that employment produces, but for livelihood, which is most families' short-term all.

Thus, people will complain about and suffer from the effects of inner-city decay while actively contributing to that decay through the work they do in insurance, banking, construction or local government. The units are invariably so large and fuzzy that the end results cannot easily be related to the part each person plays. And if 'Government funds' are involved, the lunacies may be that much more spectacular, with people's active compliance in the building of a power station that the electricity boards do not need, ships for which there are no cargoes, or aircraft practically certified as commercial failures. All these, as we know, are real examples.

The people I am talking about are no less worried than others about contemporary problems, but while things contrive to totter along from day to day it is not surprising that they remain largely aloof from moves for radical reform, concentrating instead on the worries of a real world of work and wages that they know well. There are other reasons why this should be so. One concerns the quality and availability of information about the planning and management of the modern industrial economy. If it comes from 'official' sources, as, for instance, do documents like Government White Papers and county structure plans, it will be couched in a language not spoken outside academic and bureaucratic quarters and will not, as a rule, go directly to those ordinary people whose interests are affected and who have paid for the information to be produced. Government and local-authority officials often express surprise at the lack of response they receive from the public to questionnaires, circulars and draft policy reports which have been distributed widely. The reason is simple: such material usually has a grand, thematic outlook on large-scale, broad-grain, long-term issues to which most citizens, beset by next-door, next-day problems, feel they cannot address themselves.

Filtered by the media, information generally becomes even less useful. Newspapers, with few exceptions, trivialize, miss or misinterpret vital issues. Television, because of inflexible programming, compresses them. The greatest failing in the non-specialist media, though, is to follow and reflect industrial management's inability or unwillingness to interrelate. Events are treated as if in some kind of Brownian motion – not quite random but very nearly so. The effect is to make it appear that our affairs are largely a matter of unconnected chance: major man-made disasters are presented as if they came out of the blue and had nothing to do with social and economic planning; housing and employment conditions are separated from the problems of vandalism and larceny; wars and arms deals are regarded as different stories; monetary inflation is decoupled as a subject from commodity trading.

It is a double shirking of responsibility, since the media strenuously promote and encourage – by and large – the development of that sort of global society in which it is essential to recognize and understand the linkages if democratic accountability is to be preserved and if conflict and catastrophe are to be averted. A negligent attitude towards the selection and presentation of news and information has not left people wholly ignorant of an interrelatedness in the various problems that beset them: there is a widespread feeling that the commercial and political institutions are blustering and blundering from one critical error to the next. But a majority of people are not sufficiently informed, equipped and enfranchised to ask the right questions or call for the right answers (and, indeed, to look at some of the big-picture problems as they are presented by observers like John Davoll).

Worse still: they are alienated from those who are. Consciously and unconsciously, the media strengthen popular attitudes towards radical opposition as springing from a dilettante elite whose social characteristics are actually quite similar to those found in the established institutions. A cartoon in the *Whitehaven News*, published during the first week of the Windscale inquiry, was apt. It depicted three pinstriped types standing outside the Civic Hall, scene of the hearings: their briefcases were labelled, respectively, Legal Expert, Environment Expert and Nuclear Expert. 'Pity,' one was saying to the others, 'we'll miss Wimbledon . . .'

To a certain extent, the caricature was a reasonable one. Those who came to oppose the building of THORP were not drawn from a broad cross-section of society. They were university lecturers, writers, economists, lawyers, planning consultants or professional lobbyists supported by environmental charities and educational trusts. Individually, they may have been hard-pressed for funds and physical resources with which to conduct their campaign and their case. But collectively, they possess similar skills, interests, backgrounds and aspirations to those whom they challenged. That is to say that they have been educated to a higher-than-average level, are well travelled, comprehensively interested and versed in current affairs, fonder of chess and rugby than of soccer and snooker, intellectually ahead of much of the reformist legislation on issues such as abortion, race relations and sex discrimination, and generally at ease with the world of ideas discussed in stimulating company over good food and wine.

These are stereotypes, of course. One of the best-known spokesmen for the environmentalists in Britain is a member of the darts team in a working men's club, where he smokes, drinks and speaks like the non-graduate engineer he used to be. A senior official with one of the electricity boards represented at the Windscale inquiry is, in off-guard moments, an advocate of repatriating coloured immigrants, returning women to the kitchen and cutting unemployment benefit. But even the stereotype fails to explain the *real* gulf between those who appeared at the Windscale hearings.

At an immediate level, it is the difference between optimism and pessimism. The environmentalists believe that the world is becoming a more hazardous, less satisfying, less caring place, in which the growth of nuclear-power facilities can only bring about further and faster deterioration. The pro-nuclear camp sees improvement all around, marred and postponed only by temporary and local difficulties (of which antinuclear movements are a part). The one regards technological advance as a potential cure of all our earthly ills; the other thinks of it as their cause.

It is today's version of the two-cultures division in society, with nuclear power and reprocessing as the fulcrum. On one side is the small-is-beautiful school, espousing resource-extensive and ecologically benign technologies; on the other is the much more powerful and – up to now – acceptable doctrine of energy-, capital- and machine-intensive economic development. Many people in the nuclear industry, indeed in industry at large and certainly in most Western governments, point to the indicators they see as vindicating their approach. People, they note, are living longer and less taxing lives and, moreover, living in a style freely and collectively chosen by the majority.

Sir John Hill, chairman of the UKAEA, recognized the argument in an address he gave at the beginning of 1978 to the Royal Society of Arts, Manufacturers, and Commerce in London. Why is it, he asked, 'when the technologist has delivered so much and has made a bigger increase in the wealth and living standards than has ever been achieved in human history, that he can now do no right?' Sir John went on to answer his question partially, referring to the problems of remoteness in a high-technology world, the environmental degradation caused by the motor vehicle, and the siting of industrial complexes 'which is decided not by considerations of the immediate benefit to the local community but by availability of transport, or water, or deep anchorages, or investment grants, or many other things over which the local population has no control'.

It was not surprising, he added, that pressure groups had built up to prevent the imposition of large industrial complexes on the community, or that there was an increasing suspicion of central administration: 'Some, I am sure, see nuclear power as symbolic of going further and faster down this road of bigness and remoteness and control by a bureaucracy, and oppose it on these grounds.' He had correctly identified one of the antinuclear motives – but produced no evidence to show that it is not a *justified* motive. He criticized, and rightly, television programmes which present the nuclear controversy in terms of 'goodies and baddies' – but confessed to the belief that it was impossible for the lay public to understand and evaluate information, for example, about radiological hazards.

'I hope I am not being patronising to the public intelligence on these matters. I certainly do not intend to be. The British public has far more intelligence than it is usually credited with and much more common sense than many intellectuals. But many issues can be fully understood only by people who are prepared to study the matter full-time, and even they have to rely upon experts in specialist fields.' This statement, I believe, can be faulted on two counts. In the first place, it confuses tactical and strategic aspects of the controversy: ordinary citizens, to be sure, are no more capable of running the nuclear industry in fine detail than they were of directing the troop movements in Europe during the last war. But a broadly informed public, properly consulted, might well have opposed and prevented the rearming and appeasing of Germany as – more recently – they might have vetoed a commitment to nuclear power.

In the second place, there is, again, the nanny-knows-best approach, criticized at the Windscale inquiry by Brian Wynne. The most common argument put forward to a concerned public by representatives of the nuclear industry is that 'we know what we are doing . . . but if we tried to explain our proposals, people wouldn't understand'. But highly complex information on a wide variety of subjects is continuously and competently handled by people; all that is required is that its relevance is made clear. A Cabinet Minister made responsible for the energy-conservation programme does not need to be an expert in fluidized-bed coal-combustion techniques or the construction of photovoltaic cells. In fact, it is held to be an advantage for him not to be a specialist, but to be capable of synthesizing. The economist Samuel Butler put it succinctly: 'The public may not yet know enough to be experts but they know enough to choose between them.'

This is not accepted by the industry. Peter Taylor, spokesman for the Oxford Political Ecology Research Group at the inquiry, submitted another reason – the rigid approach to decision-making on applied-science projects: we live in a society ruled by rationalists and reductionists because of our dependence on advanced technology. Our education prepares us for it, and seldom is there an opportunity for fundamentally alternative postures to be adopted:

'If one were to say "I oppose nuclear power because it threatens the very stuff of life", one would be branded as emotional and incapable of rational argument ... there is far more scope for interaction if one says "I oppose nuclear power because the disposal of long-lived actinides to geological formations of unproven permeability or long-term structural stability cannot guarantee against radiological hazards due to concentration in the food chains."

'Thus, anyone who can actually say that sentence and get the words right (they do not necessarily have to understand them) may become a front-line protagonist. The others must stay at the back and maintain a dutiful silence whilst their "champions" do battle.' It is an important distinction to make, and Taylor illustrates it by the appearance, at the hearings, of a small religious commune whose representative, Barbara Fish, made a simple appeal to the Tribunal with questions like 'Why must we always live with the blunderings of science all in the name of progress?' and 'How dare we "mere mortals" reason that our industries and refrigerators are important enough to put the balance [of nature] at stake?'

Such non-scientifically assembled questions lack credibility in a controversy dominated by economic and political 'fact'. With unconscious irony, the *Guardian* reported that Ms Fish 'who is twenty, undaunted by an audience which included eminent Q Cs, told the inquiry that, though it was a formal affair, she wished to speak informally, as it was the only way she could communicate'. The points are that she was a good deal less daunted by the circumstances than were some objectors who made highly formalized and technical submissions; and that – apart from some of her references to the works of spiritual leaders – she rested on the expression of sincere folk opinion.

Groups like the Friends of the Earth and the Conservation Society, though they contested the Windscale inquiry with the use of specialist witnesses and documentary evidence of a technical nature, draw for their support on a very large number of people like Ms Fish and her fellow communards who feel, instinctively, that 'things' are going wrong and that the move towards dependency on nuclear power is potentially the final mistake. Like one witness at the inquiry, they compare the modern industrial state to 'a driverless vehicle, with the throttle jammed open, heading along a cul de sac'. Their prophets – people such as Paul Ehrlich and Edward Goldsmith – tell them that the industrialized urban culture is close to its demise and that it will be a violent end unless it is met half way by a radical rethinking of the patterns of human settlement and behaviour.

Goldsmith, editor of the *Ecologist*, is probably Europe's apocalyptic-in-chief, and presented the Windscale inquiry with a catalogue of portents of imminent collapse. The words tumbled over themselves in a proof of evidence delivered with messianic urgency: 'The resource and energy shortages leading to the exploitation of ever less economic sources, the massive water works required to satisfy, in ever less propitious conditions, the practically insatiable demands of industry for this valuable commodity, the rising costs of feeding our massive urbanized population from a shrinking and deteriorating landbase, of pollution and of controlling it, of treating an increasingly unhealthy industrial population, of providing welfare for increasingly helpless people, and of combatting crime, delinquency, drug addition, alcoholism, and other symptoms of social breakdown, and of maintaining a socially acceptable level of employment for an expanding work force in an increasingly capital-intensive economy – all of these costs simply reflect conditions that are becoming ever less suitable for the industrial process.'

There was barely a pause for breath as he went on to conclude that courage was needed to face 'the inescapable fact that the Industrial Era is rapidly drawing to a close. The choice that faces us now [recalling his magazine's famous "Blueprint for

Survival"] is whether we are going to try to assure a relatively smooth and painless transition to a more sustainable society, or whether we are going to maintain a "business as usual" policy and thereby simply "delegate to disaster". Needless to say, the project proposed by British Nuclear Fuels can play no useful role within the former strategy . . .'

Disciple of doom and draughtsman of its dynamics, Goldsmith plainly says what many less fervent and self-confident environmentalists have thought and said over the last few years. But two difficulties confront anyone who listens carefully to what he has said. One is that his assured, almost gleeful litany of catastrophes-in-waiting strikes altogether the wrong note for the open discussion and resolution of this wide range of potential social, economic and environmental problems – especially when, as at Whitehaven, it contains unsubstantiated claims and factual inaccuracies.

The second is that his apparent cures seem worse than the illness he describes. Shortly before the inquiry opened, Goldsmith published what was set out as the successor to 'Blueprint for Survival': unlike 'Blueprint', which was the result of many authors' analysis, 'Deindustrializing Society' was all his own work. The new document begins with an examination of resource constraints and the external costs of industrial development – familiar themes and ones being seriously and extensively discussed by a great many international institutions. But it goes on to suggest some fairly astonishing reforms. In Goldsmith's brave new future, all consumer products would be banished, compulsory service in a Swiss-style militia would be introduced, women's role would revert to the orthodoxy of *Kinder–Küche–Kirche*, and everyone would be resettled in closed communities. All social services would be scrapped.

The proposals have some comic relief. Unilateral disarmament is counselled, but war, 'like all other pursuits, should be as labour intensive on as small a scale as possible'. No juke boxes and hi-fi, either: 'If people were to spend their money on hiring string quartets . . . many of our problems would be solved.' The underlying thesis is more disquieting. Welfare, for instance, would be dispensed at a village level – in Goldsmith's press-conference words – 'by the local chap who would be best qualified to decide who was in need'. His other suggestions include the setting-up of a Government department to wind up the rest of its fellow Ministries, business and industry. Travel and tourism would be stopped, modern medical health practice would be abandoned, the National Grid would be dismantled. And so on.

His views underline – by going to such extremes – the ideological and intellectual difficulties faced by the small-is-beautiful school as represented at the inquiry and elsewhere. Schumacher, for all his undoubted wisdom and clarity of thinking, shirked the ultimate responsibility of proposing just how those who, all over the world, passionately agreed with his critique of the modern industrial–economic–political system should act to bring about a betterment. But there are millions of people, many of whom belong to the antinuclear movement, who want to know. Telling them to join the Soil Association or search for the answers within their own souls, as Schumacher was apt to advise, is not enough.

Very broadly, the industrialized world is tending towards a centre-left position and striving for the economic base that will both justify and sustain that cast of social management: the European Common Market is the contemporary model. As an institution, the EEC countenances barely any of the fears expressed at the Windscale inquiry – or it purports to have catered for them. It has secretariats and bodies to look after the protection of the environment, civil liberties, employment and working conditions, the balanced mix of rural and urban settlements – an array of social, political and economic mechanisms to ensure that next year's lot will be better than it is now.

The general public may remain unconvinced by all this that genuine improvement is around the corner – but there is little sign, either from political parties or groups outside the environmental movement, of any positive challenge to the general trend. I have mentioned elsewhere that some European parties have changed or modified their specific policies and opinions on nuclear power, but the consensus is very firmly pro-nuclear still.

That leaves the opponents of nuclear power and reprocessing with three, possibly four alternatives.

One would be that they do nothing except to continue to lobby and, wherever possible, demonstrate their views, call for public inquiries, and hope – as many do – that commercial failure or a serious accident will bring the nuclear programme to a halt. Another strategy would be to attempt to politicize their cause along a front amenable to conventional parties. To achieve this, they would need to make many compromises and take on board issues of which they have little or no experience and cases they have never prosecuted. It is possible, of course, to find several causes that could be linked to the broad environmentalist overview – the alleviation of poverty, resolution of racial conflict, disarmament. But these causes are already supported, in spirit, by those who contested the Windscale expansion or who confronted the police in France. They are not institutionalized beneath a grand umbrella.

The far left shows scant interest in the environmentalist outlook, though there are strands that join the two. And one of those strands may well prove to be the most likely alternative for the antinuclear militants to adopt. No one has yet put forward a believable or attractive method for dismantling the present social machine and reassembling a more satisfactory one. Withdrawal from the system on an individual basis is an impractical proposition for comprehensive reform, and larger-scale and radical change could not be attained peaceably. So – the one road that remains open to the more determined opponents is the one that leads to classical revolutionary tactics. Most of the environmentalists will disown and disavow them; so, for a long time, will the rest of the general public; and so will those who espouse genuine anarchism, in which people might freely determine and manage their own affairs in pacific decentralized communities.

But there are others frustrated, isolated, embittered enough to make the nuclear issue a vehicle for attacking the major institutions of the state. The distinction between environmentalists-turned-extremists and those who may adapt the controversy to serve general ideological objectives is too fine to be usefully drawn. Suffice to say that a number of opponents of nuclear power in Britain and abroad are dedicated to changing far more than decisions about electrical-energy sources. There are meeting points, as well, for the fanatics of both left and right who would see merit in a Goldsmithian future but who are likely to attempt to bring it about by force and not by making endearingly eccentric pronouncements about string quartets.

Meanwhile, there is going to be an increasing number of disconcerted and concerned citizens of various orthodox political persuasion who feel that their views have not been and will not be listened to by their elected representatives. The media, for the most part, now report but still do not reflect their beliefs. What can they do? What are they likely to do? My guess is that a significant proportion may join the disaffected non-voters at elections. In Britain, the Parliamentary track record in discussing the macro-issues into which nuclear power falls is bad enough; individual Members' contributions frequently reveal in their full awfulness the narrow political judgement applied to such issues, the poor attendances at debates, and the lack of independently acquired and presented information. Most of it comes directly from one side of the lobby or the other.

Debating the Parker report in the House of Commons on 22 March 1978, the MP for Workington, Richard Page, typified much of the political feebleness that has run through the controversy. Would the proposed reprocessing plant be able to operate profitably and effectively, he asked himself aloud: 'For this we must put our trust in experts.' On radiological hazards: 'From the report, and not being an expert, *I can say with absolute confidence* that if one sat in the mud at Ravenglass, for twenty-four hours a day every day, drank 10,000 litres of water and ate thirty tons of potatoes and fifty pounds of scallops every day, one would still be within the radiological limits, and anyone who can do that deserves a medal' (emphasis added).

Again: 'I am considering the matter purely from a constituency view because my constituents and I have to live there. It is all right to have abstract debates, but I wonder how many Hon. Members have walked round Windscale and can talk about it with confidence.' He knows the answer to that: very few. And no MPs took part in the inquiry itself, except on distant sidelines. But one or two of those did at least comprehend the breadth of issues involved. Robin Cook, the MP for Edinburgh Central, had addressed himself to the controversy for a long time, and when the Windscale proposals were brought up in Parliament, was able to speak authoritatively on the questions of energy, environmental pollution and plutonium proliferation raised by the THORP application. Richard Page dealt with the last of those in the following way: 'There has been much talk about proliferation. I do not intend to become involved in that argument, but I believe in the old saying "Where there's a will, there's a way" . . .'

Ms Maureen Colquhoun, the Member for Northampton North, felt able to stand aside from the few combatants who spoke to the Commons debate in March and make, for Parliament, a penetrating observation. A decision on THORP would be about more than questions of new technology: it would be about accountability in its widest sense and about human ecology; it would also be about grass-roots democracy – and in that context, the Windscale hearings were an example 'of what we laughingly call public inquiry [in which] the objectors are either wealthy and able to afford counsel, or middle class and articulate. It is very difficult for us to gauge the working-class reaction.'

Where, asked Ms Colquhoun, were the voices of ordinary people? That was something that ought to concern Parliament. 'Are they so downtrodden, so inarticulate, so unhappy over unemployment, or so bemused by the language of the technologists, the experts, and the politicians, that they cannot use the system?' But it was not only people outside who were cynical about politicians – 'There are people inside this House who are equally cynical . . .' She went on to wonder whether the Windscale inquiry marked a turning point – the end of the middle-class pressure group – and although she was condensing a number of issues into a short submission, it seems clear that her point is similar to ones made by Peter Taylor of PERG: that cut-and-dried technical pros and cons are not what the nuclear controversy is really about, but that the issues can and should be decided as part of a great, informed debate in which all opinion is regarded with equal care.

The trouble with this perspective is that it requires a patient, cautious, tolerant and caring democracy. Not even the fiercest proponents of nuclear power would suggest that we have anything approaching that social ideal. We have a world consumed with urgency and self-interest, a world where, invariably, gross error can be vitiated only by increasing the scale and impact of the mistake. When the market for a wholly worthless product or service is saturated, simply loses interest or is closed by statute (as has happened with a variety of chemical and pharmaceutical products), then new markets are found, developed and exploited in less sophisticated parts of the world. We believe in peace but rest whole economies on the utensils of war. We champion the 'right to work' but admire and support the technologies that bring about a permanent unemployment. And we talk of free choice when millions are still, for political, cultural, educational and economic reasons, denied genuine freedom of choice. I am thinking not of Russia or Chile, Iran or Uganda, but of most of the ostensibly enlightened nations.

The supporters of THORP and of every nuclear project throughout the world do not appear to think that these contradictions have much to do with their cause. They have been able to perpetuate the notion that – for instance – energy sources and the technologies that surround them are somehow neutral. They are obviously not. Some will help people to realize their creative and manual potential to a high degree; others will hobble men and women. Some will conduce to the equitable sharing of resources; others will widen the gulf between rich and poor. Some will remove the need to perform dangerous and degrading tasks; others will introduce new hazards and even more mindless occupations. Some will enable us to make more decisions for ourselves; others will strengthen the influence of a centralized, technocratic elite.

Unfortunately for the antinuclear movement, the controversy has come at a time when self-denying principles are the least likely ones to be applied by governments. Faced with high unemployment, balance-of-payments deficits, chronic industrial disputes, the collapse of steel-making and shipbuilding, conflict in the Middle East, Rhodesia, Northern Ireland – it is scarcely surprising that the promise of cheap, clean energy sources for the near future should be embraced even more passionately now by most countries than they were twenty years ago. The fight to get an FBR programme approved will thus be even less compromising than that mounted for THORP.

It would require a conversion far more dramatic than Saul's for most countries to reject the nuclear option. The plutonium genie *is* out of the bottle – and though Senator Church when he coined the phrase was expressing more than a hint of constituency interest (he represents the only state, Idaho, that has an operational reprocessing plant in America), he is right. Whether he is also correct in saying that 'The better part of wisdom is to recognize this rudimentary truth rather than bemoan it' – implying that there can be no going back – no one can say. A large number of people do want to go, not back, but to a point at which desirable features of both past and present society might be combined. They would probably like to see a conservative attitude towards the management of resources existing alongside contemporary liberal and emancipated social policies.

Are they compatible? The first were the products of an authoritarian agrarian life; the latter those of *sectors* of our larger and older urban communities. It is relatively easy to be liberal and antinuclear if you live in parts of London, Boston or Bonn; less so if you work in industry in Sheffield, Philadelphia or Essen; and a lot less so still for people in Rio, Ibadan or Riyadh. One society's technogenic risk is another's potential route to economic uplift and social advance, and nuclear power is perhaps the most vivid exemplar of this schism. Developing nations – or those in political charge of them – are unlikely to look favourably on technologies and systems clearly labelled as being inferior by the industrialized countries.

That being so, it seems probable that the nuclear controversy will widen considerably over the next decade as more countries embrace the aspirations and acquire the know-how for development strategies based on high energy consumption. For the most reasonable guess is that countries like the USA, unless drawn into a major war or civil disturbance, will be enjoining the Third World to adopt a policy of controlled low growth and the use of appropriate technologies. America and Europe may well have gone for those strategies as well, in an effort to rediscover the formulae for economic and social stability. There may have been official encouragement for industry and population to redisperse in the wake of one collapse after another of major commercial undertakings. The West and Japan, now in grave difficulties, may find it an attractive capitalist proposition to establish a new economic order both for themselves *and* the developing countries – though that would still cast those countries in a market role rather than as communities to be left to determine their own objectives and exploit their own skills and resources.

The Chinese and the Russians can be expected to contest the issues even more vigorously than they do the present Western attitudes towards Third World development. It will be seen to be every bit as imperialistic to advocate industrial moderation and nuclear caution as it has been to sequester poor countries' resources in exchange for the consumer society's technological beads. The Russians have few doubts about the wisdom of striving for continued industrial growth, and the Chinese are newly committed to it. Nuclear power seems certain to be central to their schemes.

There will be serious setbacks for the nuclear industry and its supporters from open protest: the demonstrations at Narita airport show that there *is* a coming-together of environmentalists and other opposing forces on some issues. The violence at Tokyo's benighted second airport springs not just from the anger of dispossessed farmers, the search by radical students for an anti-Government focus, the fears by local people of noise and aerial pollution, the strictures of those who simply see Narita as an appallingly planned facility, or the feelings by Japanese observers that such a development will serve only to increase the megalopolitan scale and thus the ultimate vulnerability of high technology urban mass-management. It springs from all of them. These are precisely the kind of issues that unite the anti-nuclear movement and will continue through the controversy, in

Britain, over the FBR proposals when they are laid before the public.

This is looking to the mid-term future. More immediately, the conflict will wax and rage among the dominant countries of the West, with people painfully divided between their instincts and loyalties, their needs and their greeds, their means and their ends, their hopes and their suspicions. More than thirty years ago, Aldous Huxley made an inordinately prophetic and grim statement about nuclear power and the social context within which it would be managed, when he wrote: 'It is probable that all the world's governments will be more or less completely totalitarian even before the harnessing of atomic energy. It seems almost certain that they will be totalitarian during and after the harnessing. Only a large-scale popular movement towards de-centralization and self-help can arrest the present tendency towards statism.'

By many indicators – censorship of speech and press, imprisonment or deportation of political dissidents, curtailment of civil liberties, block-management of workforces, the reliance on a technological executive – much of Huxley's forecast is coming true. Save, that is, for his large-scale popular movement towards decentralization and self-help: anyone who supported such a movement could not, *ipso facto*, engineer its attaining those ends in a peaceable and organized way. No one will yield. Many will be fooled. Some will be vindicated. And moderation, as Raymond Kidwell, QC for the Friends of the Earth at the Windscale inquiry, said, will catch few ears. The politics of nuclear power and reprocessing will increasingly be those of disillusion, anger, frustration and mistrust. We shall all be the losers.

13 Postscript

Within thirty-six hours of publication of the Parker report, two things happened. The first was that the Japanese announced that they would curb the export of cars to Britain. Unless there was an element of double-bluff, the coincidence cannot have been a stronger confirmation for some observers that a formula had been reached to help the dying British motor trade in exchange for reprocessing Japan's accumulating stock of irradiated fuel from thermal reactors.

The second was that I started a file entitled 'reactions and developments', from which it was intended that revisions and updatings could be culled. In less than a month, up to the time of putting these words into print that is, the file is roughly five times the length of this book. A careful, consecutive collation of all the new material to be incorporated in the text would have involved the creation of a long, late academic treatise out of what began as an immediate and topical guide to the subject. And it would still be incomplete. I have chosen instead to telescope some of those reactions and developments: they illustrate how fluid and fervid the nuclear debate now is.

27 February: Letter to *The Times* from Professor Edward Radford, a witness at the inquiry from the graduate school of public health, Pittsburgh:

I challenged the Inspector's public assurances about the safety of exposures to plutonium and americium already occuring along the coast ... yet the press took no note of this confrontation, nor was there any comment that, during my five hours of cross-examination, Sir Edward Pochin did not ask me a single question, although he readily cross-examined other witnesses.

The *Whitehaven News* referred to me simply as a 'radiologist', which

I am not. Even Nigel Hawkes, whose review in the *Observer* Magazine of the issues presented at the inquiry is the best I have seen, failed to recognize the implications ... most surprising was the coverage by the *Guardian*, which prided itself for being the only national newspaper to have a reporter present at the inquiry every day. I gave my direct testimony for two hours on 28 September and was cross-examined for nearly the entire day of 29 September. There is not one word about my testimony in the *Guardian*'s summary of those two days. An omission of this kind by an American newspaper could reasonably be interpreted as deliberate ...

7 March: Leading article in the *Financial Times*:

Mr Justice Parker is to be congratulated. The Windscale inquiry, over which he presided, received evidence from nearly 150 witnesses and some 1,500 documents, many of them book-length, were submitted to it. No known aspect of the question failed to be discussed, nor can anyone claim now with any validity that the subject has not been fully aired in public. His report is a model of clarity ... It was right that a public inquiry should be held. It is right that there should now be a debate in Parliament. But the Government cannot easily duck the implications, which are that Britain needs a nuclear energy programme and should go ahead with it as speedily and safely as possible.

7 March: 'If it's inevitable, why not make it profitable?' – *Daily Mail*.

15 March: Letter to the *Guardian* from the Greenpeace Foundation:

The open-door consultative approach, so evident at the Windscale inquiry, where ... bonhomie became an institution in itself, clearly has its place, but with the odds stacked so heavily against the antinuclear lobby it is questionable whether such tactics are in the best interests of the campaign. If the report has been of any benefit to the opponents, it has drawn the various factions together and has led to a hardening of their resolve.

17 March: Two workers died and another ten were injured when a bomb went off inside the nuclear power station being built at Lemoniz, twenty miles from Bilbao in the Basque country.

Police reports said that the bomb was powerful and that the damage was substantial.

17 March: Members of Parliament were circulated with a hostile critique prepared by the Town and Country Planning Association, which detailed factual divergences between its witnesses' submissions and those published in the Parker report. It began with a note on the presentation and tone of the report.

Whatever view one might have of Mr Parker's judgement of the evidence, the tone is inappropriately subjective and personal to him and couched in terms as though British Nuclear Fuels Limited were innocent unless proved guilty, and even, vice versa, that objectors were guilty unless they could prove themselves innocent. Yet such a judicial tone was entirely inappropriate to the questions in hand. So many of the issues are not susceptible to proof or analysis on the basis of established fact, but depend for their evaluation upon an ability to balance technical and measurable fact with qualitative opinion on abstract aspects at the same time and in aggregate. The report does not consider what it *all* adds up to.

22 March: The Network for Nuclear Concern issued allegations of misrepresentation at the Windscale inquiry as summarized in the Parker report. For one, their witnesses had referred to the standards used in the USA to protect the public against radiation doses from fuel-cycle operation and to limit the environmental burdens of long-lived radioactive materials. 'These standards are twenty to sixty times stricter than those operating in the UK. NNC emphasized this contrast in standards on a number of occasions at the inquiry. No mention whatsoever is made in the Windscale inquiry report.' The Network went on to detail many other lengthy examples of misconstruction or misquotation of their case.

23 March: *Nature*, the normally conservative British science journal, published a leading article which concluded that

We do not know that Windscale will help, physically or morally, any other country to get into the weapons business. It may very well not. But it ought to be clear that discussion on proliferation issues is as yet too little advanced for such a step to be taken.

6 April: The National Radiological Protection Board advises the British Government that the safe limits for whole-body radiation doses be reduced five-fold. It recommends a maximum annual radiation dose limit of 100 millirem for a nominal life-span of seventy years: previously, the limit was set at the equivalent of 500 millirem. Government estimates for 1976 show that, for the population most at risk – those who eat fish landed from the Irish Sea off Windscale – the average dose was 85 millirem and the highest exposure was 238 millirem – two and a half times the new dose limit.

13 April: The UK Lawyers' Ecology Group complains to the International Bar Association about the Parker report. On the same day, the *Ecologist* magazine's editor, Edward Goldsmith, calls for a campaign of 'violent civil disobedience' in protest against the construction of fast-breeder reactors and the Windscale reprocessing plant.

14 April: According to a report prepared for the US House Government Operations sub-committee, leaked in Washington, the cost of nuclear power has been underestimated by as much as a factor of five. Massive public subsidies have concealed gross cost-overruns on construction and the future expense of waste-management and decommissioning of nuclear plants, which, at the end of their thirty-year lifespan, may cost more than $100 millions each to dismantle.

15 April: Letter to Penguin Books by Sir Kelvin Spencer, former Government scientist:

May I suggest that [this book] devote space to the carcinogenic effects of low-level exposures to radiation. The Parker report makes it clear that the Inspector was confused about this ... The proof of evidence given to the inquiry by Dr Alice Stewart is admirably clear to those with some scientific training. But to others, the very proper restraint in interpreting statistics, imposed on herself by Dr Stewart, leaves the layman with the wrong impression that her conclusions can be dismissed ... since she gave her evidence, the cancer statistics at the Portsmouth Naval Dockyard in New Hampshire have been partially disclosed: this disclosure adds force to Dr Stewart's evidence.

19 April: Wave-power is 'not just a boffins' dream', according to Alex Eadie, Under-Secretary of State at the Department of Energy. Attending the first public demonstration of a wave-power 'raft' designed by Sir Christopher Cockerell, inventor of the hovercraft, Eadie announced that the British Government was to allocate more than £10 million to research and development of renewable energy sources, of which wave devices would receive one quarter of the funds.

24–29 April: The *Guardian* carries a series on antinuclear protest in Germany, France, Holland, Spain, Sweden, America and Australia. The numbers estimated in these countries and elsewhere who actively oppose nuclear development – some 1,500,000 in all – represent what one of the paper's reports calls 'a mushrooming crowd . . . a grass-roots protest movement that is gaining strength'. In Australia, according to another correspondent, 40 per cent of the electorate is now opposed to the export from that country of uranium, seen as many as 'immoral and dangerous'.

28 April: Friends of the Earth in London present a detailed critique of the Parker report to the Secretary of State for the Environment. The report, they say, exhibits many examples of three major deficiencies: the omission of important material; misrepresentation of witnesses and submissions; and asymmetrical criteria of soundness 'such that substantive evidence from objectors is rejected while hypothetical notions from the applicants are embraced and endorsed'.

On the question of whether reprocessing of spent fuel is necessary at all, FoE notes the Inspector's insistence that, regardless of the state of our technical knowledge, it is desirable that a decision should be taken at once on which course should be pursued. 'This all-or-nothing, now-or-never approach leads him to conclude that it is not advisable to develop prolonged storage [of unreprocessed fuel] . . . he goes on to say that, if reprocessing is ever to take place, "it is preferable to start without delay . . . this is to the benefit of workers, public, and future generations alike". On the contrary, a commitment now to THORP – excluding work on prolonged storage – will increase the likelihood that if THORP, like its forerunners, does not work satisfactorily, *there will be no feasible alternative of secure storage of spent fuel*. To insist that we now place all our eggs in the untried basket of THORP can hardly be regarded as "to the benefit of workers, public, and future generations alike".'

According to FoE, the Parker findings are coloured by a misconception that appears throughout the inquiry report: that reprocessing somehow makes material disappear and disposes of toxic wastes. '[Parker] does not, however, seem to recognize that if the stocks of AGR fuel are reprocessed, stocks of low-, medium- and high-level solid and liquid radioactive waste and of plutonium will likewise "continue to build up and will have to be stored until disposed of in some manner". Reprocessing does not make spent fuel vanish. It merely divides up the problem into several new ones, some of which are at least as intractable as the original one . . . the longer reprocessing is delayed, the lower the level of radioactivity in the spent fuel, and the smaller the quantities emitted during reprocessing or remaining to be stored. Accordingly, the Inspector is simply wrong to conclude that reprocessing, if done at all, ought to be undertaken at once. On the contrary, even if reprocessing is eventually indicated, a delay will make the reprocessing easier and safer.'

At the inquiry, FoE had challenged the BNFL view that oxide reprocessing, by recovering uranium and plutonium, would make a valuable contribution to the conservation of fuel and energy. In their analysis of the report, they note the exhaustive testimony presented on their behalf by Gerald Leach of the International Institute for Environment and Development: he had submitted quantified and thorough data to show that THORP would be irrelevant to any future problems of energy supply or predictions of demand. Nowhere in the report had the Inspector seen fit to mention Leach or his submission.

'Instead, the Inspector indulges in an extraordinary outburst of obscurantism: ". . . it is clear that such evidence fell far short of what I would require were it for me to make a definitive forecast [of energy demands]. I have not regarded it as any part of my task to do so. It would serve no useful purpose, for, no matter how much evidence had been tendered, *any forecast which I might make would be as uncertain as any other forecast.*" He then, nevertheless, and apparently without even realizing it, proceeds to make just such a forecast by implication: "With so many uncertainties, the only prudent course is to adopt a strategy which will give the greatest assurance that, no matter how the variables change, the energy needed to support an acceptable society can be provided." '

As far as the economics of reprocessing is concerned, the inquiry report is accused of engaging in 'a nuclear version of the old "find the lady" street-corner enterprise' – in other words, of card-sharping. In spite of all the quantitative evidence to the contrary, 'the Inspector seems to insist that the enterprise will justify itself on economic terms, even after he himself has expressed unambiguous doubts'.

A lengthy section of the FoE critique deals with the matter of plutonium proliferation. In his report, Parker had said (of safeguards against the misuse of fissile materials): 'This system was acknowledged by everyone to be in need of strengthening and improvement . . . It is sufficient to say that it could and should be improved.' On the contrary, says FoE, 'It is *not* sufficient. FoE argued that the safeguards system could not be improved enough to cope with an international commerce in separated plutonium; and FoE adduced a wide variety of detailed expert evidence to support this contention. If it were sufficient to say that the safeguards system "could and should be improved" the world's major industrial powers would not have agreed, a month before the inquiry began, to carry out a two-year International Nuclear Fuel Cycle Evaluation (INFCE) programme into how safeguards systems could and should be improved. The inspector's casual dismissal of this key issue simply will not do.'

FoE take up the report's assertion that, since THORP would not be operational for ten years, ' "The effective risk would thus be a risk of increase proliferation, at the earliest, in ten years' time." This view is untenable. The effective risk is the risk posed by other, nominally civil reprocessing activities, which may be operational much sooner, not least that of Japan, which will be given the blessing of the British Government should THORP be approved. It may be noted without further comment that the Chasma reprocessing plant in Pakistan is now well under construction. Pakistan has only one small nuclear power station.'

The critique concludes with remarks likely to be endorsed by all the objectors to THORP and, indeed, by those who have reservations about the system, in Britain, for resolving such matters in the best interests of all concerned:

'For all its shortcomings, the Windscale inquiry was undoubtedly a landmark in British nuclear policy-making. It set standards for the public treatment of nuclear issues that are higher than those reached anywhere else in the world. It is thus all the more tragic that the hasty and erratic judgements of the report on the inquiry did no justice to the proceedings. The result has been a polarization rather than a moderation of the debate. The omissions, misrepresentations, and distortions contained in the report and extensively documented here and elsewhere have inevitably cast doubt on the value of reasoned argument. It is difficult to see how public confidence in the worth of participation in such lengthy and costly procedures can easily be restored. The Parker report has done little to reduce public concern over the wisdom of reprocessing and much to strengthen the hand of those opponents who prefer action to argument.'

29 April: At a mass antinuclear rally in Trafalgar Square, London, some of those opponents speak in support of non-violent civil disobedience. A crowd of more than 10,000 is urged to contemplate picketing and occupation of nuclear facilities if the Windscale expansion is sanctioned by Parliament and existing plans for nuclear-power generation were continued. It is the beginning of a week of international protest, including –

30 April: A mass demonstration at the Barnwell reprocessing plant in South Carolina, resulting in the arrest of more than 250 people belonging to the Palmetto Alliance.

The Swedish Atomic Forum publishes details of a study showing that spent nuclear fuel could be stored, unreprocessed, for at least 5,000 years in specially designed ceramic containers – buttressing strongly the arguments advanced at the Windscale inquiry by Friends of the Earth.

1–3 May: Similar protests in Holland, Germany and Japan.

2 May: 'Critical geological stumbling blocks' must be removed before the US can be sure that permanent underground storage sites for its atomic waste are safe, according to a US Geological Survey report. Dr Newell Trask, a member of the Survey team, tells a news conference that continued expansion of the nuclear power programme would depend on 'an acceptable solution to the problem of how to isolate from the environment for thousands to hundreds of thousands of years the current and future wastes that will be generated'.

4 May: A *Financial Times* report describes how environmentalist opposition has blocked the mining of uranium ore in Sweden. Though the industry there is said to be biding its time until public opinion has changed in favour of nuclear projects, the economics of yellowcake extraction may prove the greater disincentive: 'The cost of producing uranium [from Swedish shales] is understood to be about $35 a pound. Only a very bold gambler would bet on uranium continuing to demand its present price of over $40 a pound.'

6–7 May: 4,000 people from Britain and overseas occupy the site of Scotland's next planned AGR, at Torness in East Lothian. More than twenty groups put their signatures to a Declaration, unveiled after a weekend of speeches, displays of alternative-technology projects and theatre workshop. The Torness Declaration demands an immediate and unqualified cessation of work on nuclear projects and a wholesale diversion of research and development funds into the use of environmentally benign and renewable sources of energy such as those from wave, wind and solar power. This, the Declaration says, would provide socially useful work for all in energy and other fields.

Robin Cook, MP for Edinburgh Central, addressing the Torness protestors, tells them that £145 million has been wasted in Britain on a programme of advanced gas-cooled reactors: the programme has been an outstanding technical failure and demonstrates that nuclear power can be regarded as neither economic nor safe – adding that, if the country could not free itself of the momentum which alone kept the nuclear industry going, then, far from stimulating employment and prosperity, the industry would be responsible for substantial social and financial stress. The public would lose as governments threw good money after bad into the nuclear programme; looking farther ahead, however, manufacturers would find in the final reckoning that there was no market for the sale of equipment to the nuclear-generating industry. The combined forces of protest and economic reality would have put an end to it.

9 May: The British Government formally accepts the recommendations contained in the Parker inquiry report. BNFL hails the acceptance 'with great enthusiasm'.

15 May: The Windscale proposals are approved in Parliament by a majority of 144 after a debate in which most of the main arguments are rehearsed but at which few minds are changed. David Steel, Leader of the Liberal Party and opponent of the plan for THORP, recalls recent environmental catastrophes – the Seveso chemical-plant accident in Italy, the crash-landing of a nuclear-powered Russian satellite in Canada, and the wreckage of the *Amoco Cadiz* (from which 230,000 tonnes of oil were spilled along the French coast). Says Steel: 'Increasingly, people are asking what it is that we are doing to a world of which we are only temporary trustees. I believe the magnitude of the decision that we are asked to take today is greater than any of the examples I have given.'

More statistics are thrown into the debate. Steel mentions that the House was told in 1962 that Concorde would cost about £150 million: the latest figure available was £1,137 million, not counting operating losses. Michael Heseltine, in qualified support for the Windscale expansion, observes that 53,000 coalminers have been killed in Britain since the turn of the century. Robin Cook cites the seventeen Windscale inquiry witnesses who, in a letter to *The Times* three weeks earlier, said: ' "We each consider that our evidence has been misunderstood, misrepresented, distorted, or ignored".'

The Debate is wound up by Tony Benn, Secretary of State for Energy, who defends his Government's record on the Windscale affair. Cataloguing the processes of consultation and discussion that have taken place since the application first surfaced, he says: 'I do not honestly believe that even the sternest critic of what we are proposing ... could deny that we have opened it up, that we have encouraged discussion, and that many other bodies besides the House of Commons ... have all joined in the debate.'

On the other hand, says Benn, he fully understands the disappointment of environmentalists that their view had not prevailed. Though the decision would be made by Parliament in a few moments, 'The issues which have been raised by the environmentalists will, in my judgement, always be upon the agenda.'

During the debate, Peter Shore, Secretary of State for the Environment, announced the setting-up of the long-promised Radioactive Waste Management Advisory Committee. Its chairman was named as Sir Denys Wilkinson, a Fellow of the Royal Society, a nuclear physicist and currently Vice-Chancellor of the University of Sussex.

16 May: While BNFL prepares to formalize the signing of its contract to take spent nuclear fuels from Japan, FoE announces that it will continue its campaign. Leo Abse, MP, an emphatic opponent of reprocessing, says the fight is only just beginning. 'I don't think anyone who listened to the debate [the previous day] would doubt that, if they won the vote, they did not win the debate.'

Yet the *Guardian*, often and widely regarded as the environmentalists' voice, opines that 'public opinion, better informed as a result of the Parker inquiry, will not allow its elected representatives to sanction an unsafe venture ... what was a proper caution could be unreasonable timidity.'